Patrick Moore's Practical Astronomy Series

Other titles in this series

Telescopes and Techniques (2nd Edn.)
Chris Kitchin

The Art and Science of CCD Astronomy
David Ratledge (Ed.)

The Observer's Year (Second Edition)
Patrick Moore

Seeing Stars
Chris Kitchin and Robert W. Forrest

Photo-guide to the Constellations
Chris Kitchin

The Sun in Eclipse
Michael Maunder and Patrick Moore

Software and Data for Practical Astronomers
David Ratledge

Amateur Telescope Making
Stephen F. Tonkin (Ed.)

Observing Meteors, Comets, Supernovae and other Transient Phenomena
Neil Bone

Astronomical Equipment for Amateurs
Martin Mobberley

Transit: When Planets Cross the Sun
Michael Maunder and Patrick Moore

Practical Astrophotography
Jeffrey R. Charles

Observing the Moon
Peter T. Wlasuk

Deep-Sky Observing
Steven R. Coe

AstroFAQs
Stephen Tonkin

The Deep-Sky Observer's Year
Grant Privett and Paul Parsons

Field Guide to the Deep Sky Objects
Mike Inglis

Choosing and Using a Schmidt-Cassegrain Telescope
Rod Mollise

Astronomy with Small Telescopes
Stephen F. Tonkin (Ed.)

Solar Observing Techniques
Chris Kitchin

Light Pollution
Bob Mizon

Using the Meade ETX
Mike Weasner

Practical Amateur Spectroscopy
Stephen F. Tonkin (Ed.)

More Small Astronomical Observatories
Patrick Moore (Ed.)

Observer's Guide to Stellar Evolution
Mike Inglis

How to Observe the Sun Safely
Lee Macdonald

The Practical Astronomer's Deep-Sky Companion
Jess K. Gilmour

Observing Comets
Nick James and Gerald North

Observing Variable Stars
Gerry A. Good

Visual Astronomy in the Suburbs
Antony Cooke

Astronomy of the Milky Way: The Observer's Guide to the Northern and Southern Milky Way (2 volumes)
Mike Inglis

The NexStar User's Guide
Michael W. Swanson

Observing Binary and Double Stars
Bob Argyle (Ed.)

Navigating the Night Sky
Guilherme de Almeida

The New Amateur Astronomer
Martin Mobberley

Care of Astronomical Telescopes and Accessories
Martin Mobberley

Astronomy with a Home Computer
Neale Monks

Visual Astronomy Under Dark Skies

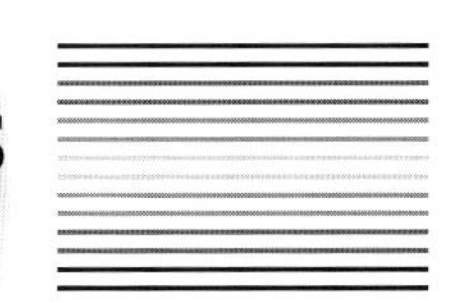

A New Approach to Observing Deep Space

Antony Cooke

With 117 Figures

Cover illustration: An evening desert scene rising into a Milky Way view, with images superimposed of a JMI NGT18 Telescope and Collins I3 Image Intensifier.

British Library Cataloguing in Publication Data
Cooke, Antony, 1948–
Visual astronomy under dark skies: a new approach to observing deep space. – (Patrick Moore's practical astronomy series)
1. Astronomy–Observations 2. Astronomy–Techniques
3. Telescopes–Technological innovations
I. Title
522
ISBN 1852339012

Library of Congress Control Number: 2005924727

Patrick Moore's Astronomy Series ISSN 1617-7185
ISBN 10: 1-85233-901-2 Printed on acid-free paper.
ISBN 13: 978-185233-901-2

Printed in the United States of America. (EXP/EB)

9 8 7 6 5 4 3 2 1 SPIN 10993101

Springer Science+Business Media
springeronline.com

Foreword

It is safe to say that virtually every amateur astronomer desires dark sky conditions to observe deep sky objects. For most of us, however, this usually involves transporting our equipment and ourselves to a dark sky site. This book, which complements Tony's first book, *Visual Astronomy in the Suburbs*, provides practical insight into deep sky observing in dark sky conditions and describes how image intensifiers and video cameras can further enhance this experience.

I will never forget my first night at a large telescope under remarkably dark skies. I was doing developmental testing with the I Cubed system (I_3) at the NASA 120 IRTF telescope facility on Mauna Kea, Hawaii. Since before dusk, I had been working inside the building with Doug Toomey, the chief telescope engineer, and at around 10:00 we stepped outside for a break. The summit of Mauna Kea, where the great telescopes reside, is above 13,000 feet (3,965 m), and there is little light from below. Hilo, the nearest city, has strict lighting regulations. Immediately upon stepping outside, I was stunned by the absolute blackness of the sky, extending to the horizon in every direction! The sky was so dark that clouds, occasionally passing overhead, could be identified only by the black shapes of sky they presented by obscuring the star fields behind them. The Milky Way, under these conditions, was a veritable fog of stars with clumpy and filamentary structures I had never seen before. I could clearly identify several Messier objects that had been invisible to the naked eye prior to this experience. I feel very fortunate to have witnessed, first hand, dark sky conditions in such an extraordinary place.

All amateur astronomers with access to an automobile and a portable telescope can experience a similar epiphany by driving to a dark sky site around new moon conditions for a night of observing. Virtually every state has numerous locations, some much better than others, that offer very dark skies (darker than visual magnitude 6 or thereabouts). Both Tony and I can testify that the results are worth the effort, even if your dark sky outings are infrequent.

The ability of an image intensifier to resolve deep sky objects is dependent on the contrast difference between the object and the sky background. As Tony's first book so aptly illustrates, a generation III image intensifier increases this contrast difference to a far greater extent than the human eye is capable of, even in less than desirable sky conditions. Dark sky conditions naturally present a greater contrast difference between the deep sky object and the sky background, allowing an image intensifier to perform at its maximum capability. A second benefit of dark sky sites involves atmospheric water vapor. At higher altitude and/or dry desert dark skies, the water column, extending to the top of the stratosphere, contains less moisture per given volume of atmosphere than a location at or near sea level with a high humidity atmosphere. The water vapor column that your telescope looks through attenuates certain optical wavelengths. The beginning of the near infrared bandwidth is between 760 and 780 nanometers and, to a lesser extent, 820–840 nm exhibits a significant attenuation dependent on water vapor entrained in the atmosphere. The gallium arsenide photo cathode in generation III image intensifiers is most sensitive to wavelengths in this region. Therefore, it stands to reason that dark sky sites that are high or dry, or both, will produce optimal results with the I Cubed intensifier. The passage of a cold front also ushers in dryer air optimizing I_3 results, even in areas of light pollution.

As a final note, I would encourage you, the reader, to become involved in one of the dark sky organizations. By preserving dark skies, we can assure that future generations of astronomers share the joy and ethereal experience of a night under a dark sky, rich with the miracles of our universe.

Bill Collins
March 2005

Acknowledgments

Special thanks to Jim Corley for his help in preparing the final version of this book.

With my telescope in the California desert

Contents

CHAPTER ONE

Introduction

Originally, when work started on this present book, I still did not have in mind that it would be a successor volume to the first, *Visual Astronomy in the Suburbs* (Springer, 2003). As things turned out, that was exactly what happened! I could not escape the fact that the new journey I had set out on was an outgrowth of the first book. However, the viewing circumstances I am concerned with in this writing, and the results that I found possible, are completely different. In this volume, I am no longer struggling to find solutions to poor viewing conditions, nor am I greatly concerned with distilling a particular viewing catalog down to just successfully observable objects.

For most of us, visits to dark sky sites are much rarer events than merely dragging a telescope out of the house. However, with some new tools recently becoming available, we now have an entire new "universe" to discover from these special places. The results are astounding, to say the least, and they certainly add fuel to the fire to justify hauling our telescopes and equipment to more favorable surroundings. (Almost needless to say, most of us do not need this additional justification!) The best part of all is that the approach I take is still contained within the "old-fashioned" spirit of real-time visual astronomy.

I think everyone can conjure up a mental image of astronomers at every level and place in history, gazing through the eyepieces of their telescopes at sights far away – true visual astronomy. We all want to connect with the cosmos in this idealistic way, but after only a few observing sessions we begin to realize that our results may not be in line with our initial expectations. Those pictures in our coffee table books do not quite match what is actually seen through the eyepiece. Much of what makes up the greater universe is to remain forever veiled from us, locked within the grasp of limited illumination. This is the challenge of seeing objects in deep space, live. The difficulty is only compounded by the difficulties of extracting fine views through an uncooperative and turbulent atmosphere.

An astronomical newcomer's entry into the hobby is likely to have been inspired by photographs taken at the great observatories, or even many contemporary amateurs' often splendid CCD images. However, in deep space at least, these spectacles have been possible only through time exposures, which often may be taken over many hours. We just have to recognize that this is the way things are; the light from these faraway places is in very short supply! Experience and acquired "seeing" techniques teach us to make out increasing detail during our viewing sessions, and it is quite remarkable how much we ultimately learn to extract from those faint ghostly images, live through the eyepiece. However, most of us would still love to be able to see those same wonders with perhaps a little more of our original expectations; certainly any novice would be well served by a more dramatically impressive learning curve. Thanks to the march of technology, now there are ways to come much closer to that original ideal. This book will show you how it is possible to attain some of the hidden potential with new image-enhancing devices, and what is reasonable to expect from using them. Though their use is not exactly a perfect science, these devices involve taking a somewhat new approach, and above all, keeping an open mind.

Of course, other than having a good telescope, we have always had the moon and much of the solar system for our viewing delight, as long as air quality allows. This has always been the case, even without our need of any special additional means. With the solar system, we may indeed fulfill our original expectations with traditional tools, and fortunately, we do not require even the darkest or purest sky for the best of the solar system to give us a good show. This holds true from almost any location. Our own "backyard" in space is readily accessible to us, although I cannot deny that a high-altitude site, in prime conditions, will indeed show us our solar system with extra-special qualities of clarity and resolution. There is nothing quite like rarified and still air, which will likely be closer to its best at ever higher altitudes. However, since exceptional viewing of the solar system is not limited exclusively to dark sky sites or high-flying locations, this volume will not concern itself with it. Approaches to viewing some of the best it has to offer were fairly extensively covered in my previous writing, *Visual Astronomy in the Suburbs*. There are countless other printed sources of information on the solar system available. Certainly, the planets radiate so much light that grasping every photon is not the issue (however, resolution is), and the electronic image-enhancing devices I espouse in this book will neither be appropriate or in any way beneficial, since resolution is mostly a matter of telescope aperture and quality. In fact, one type of enhancing device, the image intensifier, would actually be immediately ruined by such an application!

In viewing deep space, it is an entirely different matter, of course. It seems the universe guards its grandest, remotest secrets from our prying eyes quite possessively, no matter what we do. There is only so much that can register or build on the retina, no matter how much aperture we have at our disposal. (Amateur astronomers often remark on a certain disappointment at seeing a familiar deep space object live through a large professional telescope. This is not to say that the view will not be significantly better than the way they may have seen it before, but somehow it just may not seem to be proportional to the huge increase in aperture.) With the comparatively modest telescopes most amateurs use, sometimes just seeing a smudge against the background sky may be the best we can do. However a good viewing site far away from urban centers with good air will indeed transform and radically increase the potential to see ever grander sights. By this, I am referring to observing under our proverbial dark sky sites.

The many commercial telescopes available to the amateur today are surprisingly affordable. They have revolutionized what is possible to see, allowing direct viewing of many objects that were visible or resolvable only with the great telescopes of antiquity. Some of the apertures widely available to the amateur today would astonish our ancestors of only the recent past. However, there still remains an upper limit on what is possible to see in real time, and particularly, what is possible to see in a spectacular fashion. My own perpetual quest has always been to see more; if you are in any way an old hand in astronomical pursuits, it is probably yours as well.

Even in the more recent past, things have improved dramatically for us again in more ways than one. Today, utilizing CCD technology and computers, amateurs are producing images of some of the grandest celestial sights, with resolution and detail obtainable only at professional observatories until recently, and all this with far smaller telescopic apertures at that. The magnitudes of deep space objects being recorded are significantly below those obtainable with photographic film by the world's largest telescopes up till a few years ago. Some of these CCD images, made with apertures no more than 12 inches to 18 inches (30.5 cm to 45.7 cm), count among the best images of some of their subjects I have seen to date. It is no surprise that the work of the world's great telescopes has been able to advance so astonishingly in the last decade, with the use of the same or more sophisticated technology.

However, it is worth emphasizing that the magnificent results we are now accustomed to seeing in publications (even more particularly from amateur instruments) are not attainable during the actual time of observing. By this, I mean what is actually seen directly through the eyepiece, or by other means. The application of true CCD imaging is also a very sophisticated and complex process, one that many amateurs will be unwilling to learn to use. It requires considerable patience and expertise. I have to confess to being one of those still unmotivated to go this route, remaining instead firmly on course with the live visual experience. This is not to say that I do not empathize with those who do embrace the time-lapse branch of astronomy, or that I do not admire the groundbreaking achievements of their work. However, I do regret what appears to be a gradual abandonment of live observing for CCD imaging, an altogether indirect experience. I do not believe it should ever become a choice between the two; they are not interchangeable. However, since CCD imaging is not the subject of this book, nor even the implied purpose of its writing, I shall remain firmly on course for the visual approach! Does this make me an old-fashioned observer? I have to say "yes," although I do not regard my "archaic" breed as in any way missing the boat.

The human eye has a unique sensitivity to all manner of light and shade, fine detail and subtleties. It is still not possible to substitute these by any type of imaging, despite the fantastic things that CCD can indeed show us. The specific purpose behind this book, therefore, is to expand on the use of the new real-time visual technologies, so enthusiastically described in my prior volume. It is with their use that we can increase the range and effectiveness of what we can see with our own eyes in deep space, live, even from typical bad suburban conditions. Once we move to dark sky sites, these advances can go far beyond the limits previously imposed on us, and also help us better share our viewing sessions with people unfamiliar with telescopes and the skills required of them as observers. We should take advantage of our new hardware at sites that have exceptional viewing conditions. The results are not likely to be familiar to the average astronomical enthusiast.

I think it may already be clear that, in discussing image-enhancing devices in general within this particular writing, I am not referring to the specialized light filters that exploded onto the visual astronomy scene a couple of decades ago. At the time, these were quite revolutionary in the effect they had on amateur observing. They still are, of course, particularly with ongoing technical advances and the wide selections now available. These wonderful innovations have rightly become a prominent part of any visual astronomer's equipment, and I do not wish to imply that they are not significant in my own observing to this day. However, their uses have been widely described by many others before me. They featured prominently in my previous book, where they are of particular value in the bad conditions likely to be encountered in the suburbs. So do not discard your light filters in the meantime; they are no less valuable now under dark skies than they ever were! Here, many unseen, or barely seen deep space objects, particularly emission nebulae of various types, became common fare for the amateur. Even more amazingly, they work stunningly on certain things in skies impacted negatively by light pollution; such skies inspired their creation in the first place.

I also do not intend to cover in this volume, other than some of my own limited recommendations, certain additional topics, including descriptions of all types of telescopes, and their selection. These narratives are standard ingredients in many books on amateur astronomy. We do not need this all over again, along with techniques for imaging or drawing deep space objects, viewing techniques, and viewing catalogs. These topics were dealt with quite extensively in my previous book, and many other sources exist. The purpose instead is to show just what kind of results can be attained by the new breeds of electronic enhancing devices under very favorable skies. The distinction of location is in direct contrast to that of my prior book. So it is with this in mind that I hope you will forgive any unintentional void in these pages, or topics that are widely available or documented within other sources. There are already far too many books that fill up the space between their covers with identical information.

Strides in technology have given us two new, distinctly different approaches for enhancing the live visual experience: advanced image intensifiers, and advanced CCD video devices. Both of these devices compound the visuals to make a grander, more brightly illuminated and revealed whole. Recognition of their potential is still in its infancy. I hope that my previous book has helped with the acceptance of image intensifiers in this regard, as I wrote that they had actually met with complete rejection in many amateur circles. I do not believe the same is true of the less expensive, and more recently available, frame compounding CCD video cameras, though it may be too soon yet to tell. However, these new CCD video devices do indeed reveal a strong correlation with those of image intensifiers. In the succeeding chapters, for each deep space object I describe, roughly comparable results, if not of the same impact, resolution, or identical spectral emphasis, should be obtainable by either method, and so the book is quite appropriate for either.

For the record, only image intensifiers actually provide true real-time viewing, in the literal sense of the term, producing essentially the same experience as that of conventional observing. Their use is something of a revelation, and I rate them as still king for live, enhanced viewing. It seems to me that they should be the most recognized and accepted of all of the available enhancing devices! No video device yet matches it, either in resolution, or the eye's range of sensitivities to light itself. I believe that the same will still hold true to some extent even when yet more advanced video technology becomes available, because the impact or directness created by simply peering through an eyepiece is lost. However, it is also true that the differences in future video imaging, in strict visual terms, will indeed be less significant than they are today.

My own images within these pages (and descriptions of the visual impressions they made) were obtained at dark sky sites using an image intensifier system, along with matched imaging accessories and equipment in conjunction with my 18 inch (45.8 cm) reflector. For the general purpose of this writing, it would not have mattered whether these images were obtained by the method I chose, or by one of the new CCD video cameras I will describe; the parallels are often close. (Certainly, any method of video imaging produces similar resolutions; the advantages so clear in the live intensified view are lost.) Therefore, if I did not specifically mention the use of such CCD video systems for any object that is featured (since I only used the image intensifier system for the illustrations in this book), you should not take it to mean therefore that CCD video is not relevant in that instance! I hope you will treat both of these enhancing devices as in many ways similar, from the standpoint of the main attributes you will see, whether you pursue your viewing on a monitor or live through an intensifier eyepiece. There are also many other wonderful images within this book, generously supplied by John E. Cordiale, of Adirondack Video Astronomy, and Bill Collins of Collins Electro Optics. Their own contributions also provide very real impressions of what may be attained by other enhancing devices and imaging techniques, and will provide you with the widest possible comparisons for your own expectations.

In any event, the potential of what lies ahead is extraordinary. In many instances, you will be no less than astounded, but first you need to decide on the approach you will take. This will be the topic of the next chapter.

CHAPTER TWO

New Tools

Although the thrust of this book is substantially different from that in my original book, it is a logical outgrowth to expand the potential applications of image intensifiers and other enhancing devices in this latest writing. Having raised the subject so passionately in the original book, I realized that there is also a direct tie-in to the new wave of video devices just entering the market. However, because *Visual Astronomy in the Suburbs* was primarily concerned with overcoming some of the great challenges of suburban viewing, book accounts of the use of image enhancing devices (and especially image intensifiers) at dark sky locations must have seemed a tantalizing prospect to some, and the topic was left still not addressed at the conclusion of my last book. I am quite sure many readers of that volume came to realize that if what was shown in suburban skies was possible, and to that degree, then at a dark sky site the potential for the new devices would indeed be daunting, thus the focus of this writing. In many ways this present book is intended to dovetail into where the first left off, and I hope you will forgive, but understand, the occasional references to it. I hope you will also forgive some repeated information, which would render this present volume woefully incomplete to a reader who has not read the former one.

Let's set out by looking at today's cutting edge in image enhancing equipment for visual astronomy. I think it makes no sense to stop at less than the best grade of equipment you can justify or afford. Subtle differences are more critical than you may think; viewing deep space is something always at the very threshold of our vision, no matter what advantages we can give ourselves. There always seems to be tantalizing additional detail, struggling to make itself known right at the limit, and there is never a time when you feel you can comfortably see all that you wish you could. It is also necessary to break down our review of available equipment into separate categories, since the functions of the two primary types I will explore with you are ultimately so different.

Image Intensifiers

With the military origins of image intensifiers, it is one of the better fortunes of modern warfaring technology that now we have them at our disposal for other uses, including medicine and astronomy. Since they may be adapted to use in the same way as one would conventional eyepieces, they represent some of the ultimate tools for enhanced real-time viewing. Unfortunately, they are also the most expensive option. If you can obtain one of a good modern design, you are fortunate indeed, and I doubt you will ever regret your decision. These devices are among the most powerful real-time astronomical accessories to come along in years, unlocking the potential for astronomy like nothing else. You may be wondering, "If they're so good, why do not I know about them?", and indeed, that's the $64,000 question. Some of the answer may be that it has only been in recent times that they have been able to provide sufficient image quality and resolution to make them truly viable. But that cannot be all of the answer; some of it is also tied up in lingering perceptions among the amateur astronomical community.

That such wonderful innovations have occupied a no-man's land in amateur astronomy for so long is indeed perplexing, but things are slowly changing. Something else that does not help modern image intensifiers' path toward acceptance is the short shrift they have been given in the world of published books and magazines for amateur astronomers. To most people they still remain by and large an unknown or untried entity, even – if I may say it – a dirty word. Although these devices are often dismissed as being too expensive, many amateurs routinely spend much more on other accessories for their astronomical pursuits. In some cases image intensifiers even have been greeted with outright hostility. Having heard of some folks at star parties refusing even to look through a telescope equipped with one(!), I remarked in the past that this reminds me of the man standing on a street corner trying to give away money. In a strange contradiction to the relatively sorry state they still occupy in the amateur's universe, image intensifiers have been around in professional observatories for many years, and their value duly noted, even lauded. In my book, *Visual Astronomy in the Suburbs* (Springer Books, 2003), the application and special value of image intensifiers for amateur observers in unfavorable viewing locations was a central theme, albeit not the only one. The only other reference to them I can recall having seen in practical astronomy books is within the pages of *Astronomical Equipment for Amateurs*, by Martin Mobberley (remarkably also published by Springer Books). Nevertheless, despite certain practical applications for them being discussed, the range of possibilities for these amazing devices remained largely untouched within its pages. I surmise that most of this stems directly from the book's date; while this is only a little while ago, the more recent developments in image intensifiers for astronomy were too new to be much known at that time.

Long before that time, articles had occasionally appeared during the 1970s in *Sky & Telescope* magazine, detailing the use of early image intensifiers in amateur astronomy – this periodical apparently has always been open to the subject. (A recent article in the same magazine detailed the use of image intensifiers in the field of meteor observing and recording.) Although real-time viewing was a feature of some of the advertisements in the 1970s, the emphasis of the articles that appeared at the time seemed to be mostly on photographic applications, and particularly, the speed with which results could be obtained. For telescopes lacking accurate drives, or even equatorial tracking, the value was clear. Now, dramatic as the potential was then, and put forward as primary motivation for obtaining an image intensifier, it is not surprising that the advent of better telescope tracking and CCD imaging would eclipse the old image intensifier technology. As the CCD revolution came in and the first astounding results were seen, the meager attention given to image intensifiers stopped in its tracks. Not only were CCD exposures relatively fast, but their resulting astonishing imagery made other imaging systems seem irrelevant. The quality of the images obtained by the early intensifiers were hardly comparable to that which the new CCD technology could offer. They were not even any match, for that matter, to good conventional photographic time exposures of the day obtained with accurate tracking. It would seem there was no longer much reason to consider image intensifiers for the average amateur observer. The accompanying expense was the final nail in the coffin.

Only when it was appreciated that new advanced generations of image intensifiers could offer something completely unique – real-time enhanced viewing – would there be a renewed interest sparked in them, and justification for the expense of purchasing one. This unique potential, in fact, the greatest attribute that image intensifiers would hold for amateurs, remained largely ignored for years. It is still widely unrecognized. Enter Collins Electro Optics, who, only a few years ago, developed and began marketing its highly advanced and complete Generation III eyepiece intensifier – the I_3 (I Cubed) Piece, to give it its proper name. A highly credible, true real-time viewing application was the promise of this new device; here was something that lived up to the term *unique*; even professional observatories and NASA use it. Can you imagine having access to virtually the identical piece of equipment used by the pros? In a rare exception among magazines, *Sky & Telescope* printed what could only be described as a rave review of this new device when it entered the marketplace in 1999. Nevertheless, apparently this review still sparked little interest or acceptance among much of the amateur astronomical community, and it remains to this day the only device specifically made as an astronomical eyepiece. I commend the magazine, however, for trying to spread the word and for being so open to the advancing technology. The beautiful thing about this particular device is its total integration with astronomy, and while serving as an intensified eyepiece, it also can double in its role with CCD video cameras (more later). It is hardly, if at all, bigger or heavier than a modern wide-angle multielement eyepiece, such as a TeleVue Panoptic.

There is an upgraded version available now for a very reasonable extra cost, using "thin film technology," corresponding to what used to be known as a Generation IV tube. Previously, such an upgrade was prohibitive to most budgets. This improvement offers the highest resolution of any electronic image enhancing device on the market, even more than the best frame integrating video

Figure 2.1. The Collins I_3 Image Intensifier eyepiece with 1/14 inch (31.75 mm) adapter, which was used for all the illustrations in this book; pictured with optional 2 inch (51 mm) adapter (left). Attached low voltage battery power supply at right of unit.

cameras. The further enhanced performance it offers is more dramatic than I had at first realized, but it by no means relegates earlier generations to irrelevance! Even stellar points and the surrounding halos of bright ones are greatly reduced, giving the image an even more natural appearance. However, while thin film tubes should be considered as desirable, by no means are they essential for obtaining wonderful results with intensified astronomy. All of my own illustrations in this volume were made with the standard Generation III tube, as they better typify the type of result that may be expected with most equipment. (See Figure 2.1.)

Unfortunately, without special clearance, the I_3 is unavailable outside the United States, due to relatively recent export restrictions surrounding its advanced ITT intensifier tube component. Apparently, exceptions are made only to research and educational institutions within countries that enjoy good diplomatic relationships with the United States! While you may be able to connect with such an institution where you live, if obtaining access to an I_3 is not in the cards, there are still other excellent alternatives you can pursue. Collins is permitted to export the entire I_3 housing, including the mounting adapters, TeleVue eye lens component, and built-in power supply, minus, of course, the intensifier tube to interested customers. I mention this because there are some intensifier tubes by other manufacturers in the United States and other countries that will fit the I_3 assembly, as is. Consult directly with Collins as to what may be applicable. A tube manufactured in Holland by Delft Instruments is one example. Although it is only a Generation II tube, it has a fine response in the blue portion of the spectrum, and in this regard it has a unique value compared with Generation III tubes, which are geared more toward the red and infrared. The most significant downside of this particular tube and its generation (II), however, is the significantly increased signal-to-noise ratio. Carefully chosen alternative intensifier tubes, of a similar generation to the I_3, while probably not quite the equal of the ITT tubes, will still prove more than up to the task. Other parts of the entire I_3 may also lend themselves to home-built designs.

More significantly for overseas enthusiasts, moderately ingenious amateurs should be able to fashion such astronomical intensifier units from commercially available new or used components, taking the schematic drawing I have included below. This is based on the I_3 model; the concept is simple. The telescope provides the initial focal plane and image. All that is needed is the image intensifier tube itself, with a small electrical power source (separate or connected to the assembly), the focusing mount adapter, and the eye lens component to magnify and flatten the image in the (normally curved) fluorescent screen. This curvature is the normal state of affairs with screens of image intensifier tubes. In the I_3 unit, correction is accomplished by a special multi lens component made by TeleVue, something akin to a magnifying glass with field compensation. In such a home-built image intensifier eyepiece, such optical considerations will require similar solutions. In one of those amazing quirks of fate, a standard Plossl eyepiece of around 25 or 30 mm focal length will be found to flatten the image sufficiently, as well as providing the necessary magnification for fine performance. The eyepiece should be placed appropriately from the viewing screen in a mount with an adjustable position, such as a set screw, so that individual observers may bring the image on the fluorescent screen itself into focus. This secondary focal position is independent of telescope focus, and is a factor of an individual's own eyes. (See Figure 2.2.) In the Collins unit, focusing the image on the phosphor screen by the TeleVue eye lens component is accomplished by helical screw thread. This form

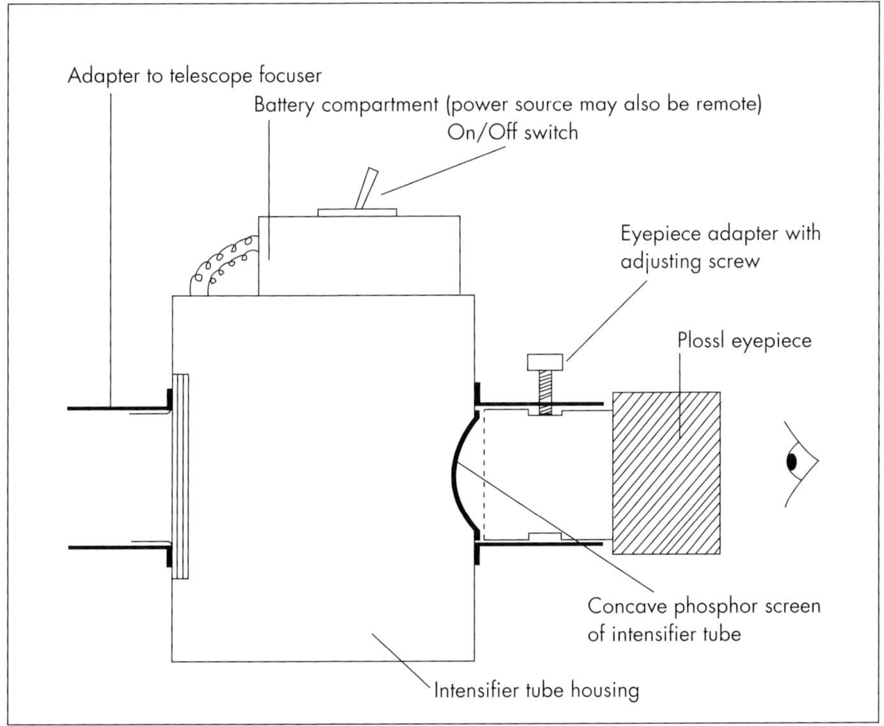

Figure 2.2. Simple configuration for building an image intensifier eyepiece.

of secondary focusing is independent of the telescope focuser, of course. See the Appendix for other manufacturers and suppliers of image intensifier tubes and equipment.

The fundamental concept is simple enough, but be sure to build the unit so that it is stable and straightforward to use. Nevertheless, it is unlikely that it will be possible to fashion a unit so exquisitely conceived and executed as the I_3, so if you can obtain or access one of these premier units, that remains my first and foremost recommendation. It is also unlikely that a new unit of equal quality could be made less expensively by anyone. However, its quality and design should remain as the role model.

Since I see myself primarily as a visual observer, for me an old-fashioned, nineteenth-century(!) viewing approach is still what my own interest in astronomy is all about. I have nothing against using all the innovations of the present day to help me in this quest, though. This originally led me to consider the use of image intensifiers when I was searching for a way to counter some of the bad suburban viewing conditions prevalent where I live. The skies around these parts (the coast of southern California) are flooded with urban light, frequent haze and marine layer, and all the joys of suburban nighttime air. I already had tried to hit the problem square on by procuring the largest readily movable telescope that I was able to handle on my own. Of course, I never anticipated results comparable to those at good dark sky sites. However, ultimately it clearly illustrated to me that, capitalizing on aperture alone, the prospect of the kind of spectacular observing I had in mind from my home site was still limited to the solar system, and no more than a handful of deep sky objects at best.

It is hardly surprising that the prospect of a device that could greatly enhance the ability to see deep space objects in real time from my home would have great appeal. After all, things were not going too well for me, as I struggled to keep the astronomy bug alive while confined to my suburban lair. Once I had determined that, in my circumstances, any and all solutions were worth looking for, I became aware, through advertisements, of the I_3 (again, in *Sky & Telescope*). This device looked interesting! So I called the phone number in the ads and asked many questions about suburban applications of this image intensifier device. After some quite encouraging answers, I decided to make the not inconsiderable financial plunge and give it a go. Once I had the new toy in my hands, there was no turning back; the results were astounding, and the unit was unbelievably user friendly. Wow! The bug not only stayed alive, but thrived!

Why had I heard or read about such prospects from so few amateur observers? (However, if you search, there are a few very favorable comments on several Web sites, i.e., http://www.scopereviews.com/page3e.html.) Could it be that there remains a certain stigma, that something so pure in concept and execution could not be taken seriously, or that some purists even think of their use as, Heaven forbid, "cheating"? It is certainly a possibility. After all, the image is an electronic one, and not "pure" in the traditional sense. Maybe some observers had experience with earlier-generation devices in unfavorable conditions; the grainy images probably resulting from such use might have put anyone off. But let's keep looking for other possible explanations. Could CCD imaging have overshadowed actual real-time observing so completely that there is no longer sufficient interest from anyone to merit much discussion on the traditional approaches of observing? Well, I hardly think so, but it's a good question. Perhaps those observers who are still passionately interested in real-time astronomy still remain unaware of the potential of modern intensified or viewing, or maybe, just do not want to know. To them, it could be just one more thing to investigate and even purchase; it would seem an expensive gamble, that's certain. I am sure that the issue of cost is a common justification for leaving them alone. However, compared to CCD applications, the comparison is not close: a quality CDD camera, accessories and processing software usually are decidedly more expensive. A multitude of amateurs apparently consider this latter type of equipment eminently affordable! However, these amateurs have apparently already abandoned much in the way of live viewing. What a pity.

I have run into the same resistance, the "immovable wall," even in many specialist telescope stores that claim to carry "everything". It is difficult to convince these astronomical equipment suppliers to stock these devices; I guess, in the circumstances, they do not have too much demand for them. The lack of acceptance by the mainstream amateur astronomical community would tend to prevent much chance of getting the good word out. Very likely you will hear damning words being spoken in many amateur circles, even by those who espouse expertise but may have never actually tried such a device! It is always possible there are those whose own experience even of modern image intensifiers has been brief and possibly unsatisfactory; image intensified observing does take a little getting used to, and some mechanical adjustments to the telescope may also be needed in some cases. But this is just the prelude to a brave new world.

Whatever telescope you have at your disposal, adding a modern image intensifier will double, triple, or even quadruple its light grasp capability. Essentially, my 18 inch (45.7 cm) telescope becomes equivalent to something like a 50 inch (1.27 m) or more. With the upgraded thin film intensifier tube, the telescope performs more like a far grander telescope even than this, maybe by several times. Imagine having the personal full-time use of an aperture not much less than some professional observatories! Now we are going to use it at a dark sky site such as these observatories usually enjoy. But your expectations will need to be tempered according to telescope size and the effective magnification that results from it. However, your 12 inch (30.5 cm) could equal a 30+ inch (76.2 cm+); your 8 inch (20.3 cm), a 20+ inch (50.8 cm+), and so on. None of these effective apertures is anything to sneeze at. The lower magnifications and wider fields that many smaller apertures will produce are also a factor in the live viewing of any object, since the images will be correspondingly brighter at lower powers with any given aperture. This effect of magnification can indeed compensate, to some degree, for a smaller telescope than that which one might ideally choose or have available. The same degree of detail will, of course, not necessarily be present because of image scale and lower resolution.

We must, however, always bear in mind that image intensifiers (or, indeed, any type of enhancing device) do not work equally well on all things. Therefore the gain we may hope for will be true only for the objects on which the type of intensification of any particular tube is most effective. However, there are so many such objects that respond favorably that the results do indeed justify the claims of greatly increased effective aperture. Suddenly, the cost of an intensifier seems trivial! It's a lot less than a comparably performing scope, in some cases a whole bank vault less! Remember though, that we are referring to the benefits of increased light grasp only, the most significant ingredient for seeing objects in deep space, live. This becomes quite obvious if you look at a really fine deep space CCD image taken through, say, a 16 inch (40.6 cm) telescope, and compare it with an older photograph from the Mount Palomar 200 inch (5.1 m). Nothing can increase the resolution itself of a given aperture, though this is far less of a factor than when viewing or imaging the solar system. In real astronomical life, the larger the aperture, the less the telescope is able to realize its theoretical resolution anyway, so we are losing less than you would think in visual performance in a direct comparison. (This resolution limit is due to atmospheric disturbances across the ever broader wave front, which are responsible for blurring whatever is in the telescope's view. The development of adaptive optics at professional facilities speaks to the extent of the loss of the potential resolution, and the need to find a means to address it.)

When seeing the image through an image intensifier, all wavelengths of light are absorbed by the unit, and new illumination is generated in one frequency only – green. All image intensifiers work by receiving photons (of light), triggering multitudes of electrons, which in turn illuminate the internal phosphor screen as a tiny green image. The actual increase in illumination can be up to the order of 50,000X in a Generation III tube, but the exact amplification will depend on the nature of the source. In any event, the eye will not perceive the enhancement to be anything like this degree. It is something akin to increasing apertures; once one has observed with moderately large telescope sizes (say, 12 inches to 24 inches/30.5 cm–61 cm) the eye is not necessarily overwhelmed by the increased brightness of images seen through ever greater telescopes. Nevertheless, the degree of image enhancement is there, and inexperienced observers will find the images infinitely easier to appreciate than those in conventional viewing. While we may be used to maximizing our own astronomical eyesight, the uninitiated viewer often struggles to see any of the subtleties in the nonintensified image so readily apparent to those of us who have spent years looking at the stars.

The images we see therefore are electronic, highly illuminated manifestations of newly generated light, seen via a tube that has a phosphor screen. So what we see ultimately approximates black-and-white viewing, albeit actually in shades of green. In fact, the color of the phosphor screen was selected by the designers of intensifier tubes because it best matches the most favorable average response of the human eye. In any event, there is little color visible to the human eye in deep space subjects, and with conventional viewing, the increasing effects of dark adaptation drastically reduce any awareness of color anyway. With image intensified viewing, such limited dark adaptation as we are able to obtain, will, over time, create something of the illusion of black-and-white imagery; the green color seems the last thing on our minds. Since most deep space objects exhibit little or no color in live viewing, other than shades of white or gray, this is not much of an issue. (See Figure 2.3.)

We've all seen the images of military operations after dark, recorded by TV cameras using image intensifiers. By the grainy and snowy appearance of many of them, it seems that not all of those images were made using the most advanced intensifier tubes, as not all forms are created equal. Today's Generation III and the even more advanced thin film image intensifiers represent the state-of-the-art

Figure 2.3. M13 in Hercules; 6 second digital exposure with 7 inch (178 mm) Astro-Physics refractor and I_3, taken in suburban Denver. The image actually reproduces much of the impact and brilliance of the live view from dark sky sites successfully. (Courtesy W.J. Collins.)

in a long history of refinement. Not surprisingly, they were developed for military use. Meanwhile, do not confuse image intensifiers, especially these advanced varieties, with night vision devices, also used to show military operations. These beam out infrared light from their own source to make objects visible, but of course would have no value in astronomy, for obvious reasons.

A degree of electronic "noise" and "scintillation" (together, these look like subtle sparkling and moving speckles) are also part and parcel of the function of image intensifiers. The much greater degree of it, and the fuzzier definition, that the older generations of image intensifiers exhibit, may also have given all intensifier technology a certain reputation. This might well have influenced some people in the negative opinion they hold of them. The problem is only increased as one's location moves closer to urban centers, with the growing inherent light and sky pollution. In a Generation III unit, the superior results, compared to earlier generations, are partially due to the much higher signal-to-noise ratio of the design. The so-called Generation IV (thin film) tubes are even better in this regard.

If we consider just the simple mechanics involved, just bringing the image of an image intensifier or camera to focus obviously is different from that of conventional eyepieces. In an intensifier, because the final image we see is a combination of two separate optical paths, the position that the focus may occupy could be very different from that of our conventional eyepieces. In a few cases, some further simple modifications to your telescope may be necessary as well, in order to give the focusing unit itself sufficient travel to allow for viewing the image. Fortunately, and quite conveniently, in another one of those little quirks of fate, the use of a Barlow lens will overcome the problem in almost every instance. The image scale such configurations produce may also suit your live viewing better with many subjects. The I_3 may be supplied in a configuration specially tailored to a given telescope, but would need specific information about the telescope's design. This may be useful if you intend to use it without the Barlow addition. Another focusing issue with the I_3: although it comes with its own built-in power supply (a surprisingly small standard 3 volt lithium battery), it is contained in a compartment on the outside of the intensifier component itself. Depending on the way you intend to position the I_3 in the focuser, its placement on the device may not allow the intensifier to be positioned sufficiently deep into the focal plane. In this case, the device is also available with the power supply detached from the main unit, supplying the electrical current by wires. Similarly, in an intensifying unit of your own design, you will need to consider all factors concerning the travel of the focusing mount in your own particular telescope.

In something of a throwback to the past regarding the ease of obtaining quick photographic exposures with image intensifiers, this particular attribute is valid once again for amateurs with telescopes lacking tracking capabilities. A Dobsonian mounted reflector, without any tracking capability, can deliver excellent results if the user works quickly and knowledgeably. For those with only roughly aligned equatorial mountings, accurate tracking is also hardly an issue, especially with the moving images of CCD video (more below). A very sizable Dobsonian, such as 24 inch (61 cm) or larger, equipped with a Collins unit, would cost a fraction of a substantially smaller catadioptric with all the gadgets and features known to humanity, as well as being complemented with full CCD capability. Therefore, what an option this presents us! My own telescope, a JMI NGT-18, an 18 inch (45.7 cm), F4.5 equatorial Newtonian, equipped with my intensifier, would seem to combine more flexibility still, since it has excellent tracking capabilities. Just by itself, the telescope performs amazingly. Used in conjunction with only visual applications of the image intensifier, it is still a less expensive option than if fitted with good CCD equipment. As it is, in my own situation, I cannot pretend to have economized with all the various accessories I have added in the meantime! However, I have not lost the live visual capability I value the most. (Believe it or not, if it were actually feasible for one person to move an even larger telescope, I would still be thirsting for even greater apertures, the common manifestation of "aperture fever"!)

Allowing for the immediate differences of image intensifiers in general (green color, scintillation, less pinpoint fidelity, etc.), the image will then appear similar to looking through a conventional eyepiece. One other key difference is that intensified images do not have the sense of depth that a conventional view has, appearing somewhat one dimensional by comparison. The field of view obtainable on the tube screen will also not be as wide as those of modern eyepieces, but not at all dissimilar to the fields of view obtained with the standard Huygenian or Ramsden eyepieces widely used just a few years ago. Remember, we are not looking straight through a conventional optical system, but peering into an electronic one. The optical path is quite different to that of a regular eyepiece, as the actual light forming the image is not that which originally entered the telescope. It remains one of those amazing gifts of good fortune that the real-time experience is faithfully preserved in the image we are seeing, as well as the subtleties the eye detects so well.

Since every type of image intensifier is a wide variable, along with its manufacturer and generation, it is impossible to provide exact performance expectations for you. And of course the telescope used, in conjunction with the intensifier, as well as the sky conditions themselves, play an equal, if not more important, role. You should use the illustrations in this book as a good representative starting point. While bearing in mind that I have used a sizable telescope in good dark skies, along with the best intensifier available, I would stress that the results I describe and illustrate here are not meant to imply that your own hopes may not be realistic, given less favorable circumstances. Remember, for example, a smaller aperture of similar focal ratio means less light grasp and less magnification, but also may mean as bright an image. This effectively cancels some of the advantage of illumination, at least, of the larger telescope. Then consider that pictures of the type I have selected still fall far short of the live experience itself. The images made via the most advanced I_3 unit available are still not capable of equaling what you will see with your own eyes, live through an enhancing device, with substantially less means available to you than that I have been fortunate to have and to illustrate this book.

A good reference for optimum magnifying power can be found using the I_3 as a model; it can be supplied with one of two available effective focal lengths, 15 mm and 25 mm. These are designed to produce equivalent image scales of eyepieces of similar focal lengths. I recommend the 25 mm version, because the magnifying power is achieved only by magnifying the intensifier screen itself. This is also my recommendation for a unit of your own making. Magnification takes place after the telescopic optical image is projected onto this screen. The same screen is used in both I_3 versions, so therefore the higher-powered version also increases the scale of the electronic noise; it is only the screen size we are amplifying. Because viewing deep space objects is a primary reason for using an intensifier, really high powers often are not ideal; almost all of my own viewing tends to be at powers from around 75X–150X, yielding an acceptable image scale and brightness. For small subjects that are bright enough, I can still readily increase the image size again using ever more powerful Barlows, and now the image noise is not increased, because we are amplifying the scale before it enters the intensifier! Actually, on most objects, the use of much higher powers than these with all but the largest telescopes will result in too dim an image, and maybe not even a visible image at that. Because brightness, as opposed to magnification, is the main issue for most intensified deep space viewing, it is darkness and clarity of air, more than steadiness, that we need. Our subjects seldom require the most stable air, or the high level of resolution such air allows.

The image that illuminates the phosphor screen of an image intensifier will also be upright, as in terrestrial telescopes. In astronomical telescopes, after all, the inverted image is only a result of using fewer optical components, in order to conserve light and its subsequent degradation. With an upright image, only some small adjustment by the experienced astronomical observer will be necessary, because depending of the position of the eyepiece and the telescope, up or down may well be moot anyway. Only in a direct comparison with conventional viewing is there really any kind of issue, although most would agree this does not amount to much. We require, however, much less of a mental adjustment than with the lateral inversions (side to side) certain star diagonals impose on the image with catadioptric telescope systems popular today. In any event, star charts are usually printed noninverted.

Any or all of these newly experienced factors, unique to image intensifiers, might seem to be a distraction to the user. Surprisingly, once the brain becomes accustomed to the differences between intensified viewing and the conventional approach, the images appear more normal; even the snow is discounted. Intensifiers are very nice to use, and it turns out that they are remarkably friendly devices. There is no tinkering of controls to view the image; just point and look! The trick with any image enhancing device is to learn the subjects that will likely benefit the most, and how to see the full extent of what is revealed in the image. This is a similar process to "learning to see" with conventional eyepieces, but there are significant differences. Now the image is much brighter, but the character of the image is often quite different. My initial experiences with an image intensifier resulted often in missing the most striking details being revealed to me, live, for the first time! However, even though the images are so bright and readily accessible compared to normal viewing, the intensity of the output will drastically reduce dark adaptation of the eye. The amount of adaptation you will attain nevertheless is quite sufficient for this type of observing, though. Any time a faint astronomical object is made into one that is relatively brightly illuminated on any type of screen, internal or external, limited dark adaptation will be the result.

For really spread out or diffuse subjects, Collins also supplies various lens attachments, at different prices, that fit onto the image intensifier and are essentially video camera lenses with variable irises. These are also available for overseas shipment, but it should not be too difficult to adapt another type of camera lens (one with focusing and a variable iris) for the same purpose, in conjunction with other intensifier units. The one I selected is probably the most popular model from Collins: the 50 mm F1.3 lens (Figure 2.4). It produces a view through the image intensifier eyepiece component corresponding to no magnification at all, allowing, among other things, some truly spectacular viewing of the great star fields in our galaxy, as well as certain spread-out nebulae that might otherwise be neglected. You should not hesitate to obtain one of these, or something similar, for your field trips; the views obtainable are among the most spectacular you will ever see live. There is no way to represent in the illustrations within this book the full impact of the kind of viewing produced in this manner, but rest assured, it is more than worth the relatively small expense to have such capability. A richest field telescope, through which we might also observe the heavens using an image intensifier, would also serve us well in a similar capacity, although it will produce some magnification.

Some important words here about the various generations of image intensifier, featured so regularly in my descriptions: while I believe all generations of image intensifier have potential value when properly applied to astronomical pursuits, Generation III units offer the most value and potential overall, providing an

Figure 2.4. The Collins 50 mm F1.3 lens attached to the I_3.

excellent response across valuable portions of the spectrum, and contrast against the night sky with minimum noise. Should you go to the trouble of assembling your own intensifier unit, I would not advise using Generation I tubes. These tubes (still available, and much less costly), with their far lower signal boost (apparently of the order of only around 1,000X), and also hugely inferior signal-to-noise ratio, permit only very limited real-time use on only the brightest objects. Generation II units are significantly better, and would be a recommended minimum for live viewing purposes. Remember, they also have a response more skewed toward the blue portion of the spectrum, so they have a unique value all their own. Closer in design to Generation III intensifier tubes, Generation II tubes nevertheless carry with them greater electronic noise, resulting in reduced contrast between the background sky and the deep space object being viewed. So they remain just that, a minimum requirement. In that these earlier generations (I and II) are also less responsive to red and infrared wavelengths, they may also be less valuable in today's commonly imperfect skies. So for the best results, our choices are not very wide. Dark sky sites, though not all created equal, will offer the best results compared to other locations. Lack of skyglow is a significant factor in the performance of these devices. While the Generation III intensifier thin film tube is best, high-quality Generation II intensifiers can still provide fair results when used away from skyglow.

It is true that the spectrum of light forming the intensified image is somewhat different from the eye's natural spectral response. Image intensifiers respond to a wider range of wavelengths, and also amplify parts of the spectrum disproportionately. This is also true of the CCD video cameras described in the next section to some degree. However, far from this being a disadvantage, it provides another illustration of the unique value these enhancing devices have: we are able to see subjects, and interesting aspects of them that might otherwise elude us. Generation III devices provide another great advantage, since they are highly responsive to the red and infrared portion of the spectrum; it is within these wavelengths that many of the wonders of deep space lie. These rich spectrums of light wavelengths are, fortunately, utterly different from that of skyglow. In many instances from urban areas this particular frequency actually enables us to see otherwise near-invisible objects. Here, if all wavelengths were amplified equally to the skyglow, we would probably not gain much advantage. It is precisely because the intensifier amplifies wavelengths not so well perceived by the eye that the results are so striking. However, skyglow will also be enhanced to some degree, producing a green hue to it as well. The city dweller will find this background color to be not as pleasing or effective as the backdrop provided by a true dark sky. However, satisfactory viewing is still possible, especially with Generation III intensifier units, because of the greater enhancement of the other wavelengths from deep space. Considering a city dweller's circumstances, he or she will probably find the green background more than acceptable, given the advantages of an intensifier system. I should say, though, that within the confines of urban living, or even nearby, use of a Barlow lens darkens the background skyglow so efficiently that a boost of only 2X almost eliminates the green background, sufficient for most viewing applications. Therefore, at my home base I hardly ever use my intensifier without the 2X Barlow. Any more power usually darkens the subject image too much.

Being based in the suburbs, like most amateur observers, I am all too infrequently able to access dark sky sites, so my image intensifier must especially be given its due. It is sufficient to say that for me, at least, it saved the hobby as a regular activity. Any of the enhancing devices I describe in these pages can probably do the same for you as well. However, little did I consider at the time of my original quest that my I_3 would be of such great service at dark sky sites as well.

See the Appendix, "Understanding Image Intensifier Tube Performance Specifications," for further information regarding the above section.

Video Devices

If the very special advantage that image intensifiers offer is nevertheless a too costly option – namely, that of true real-time extraordinary viewing – there are indeed some cheaper alternatives to use with our telescopes: frame integrating video cameras. However, we must settle for viewing images on a monitor instead of the live view through an intensifier eyepiece. This results in something less in immediate impact and crispness, although, since these alternatives work amazingly well and are a much less costly option, they are certainly highly viable. Does the considerable difference in cost between most of these video systems and image-intensified devices still make sense in choosing an image intensifier if possible? Definitely; I would still say so strongly if you can afford it. The image intensifier provides a live "eyepiece experience" not possible to duplicate by any other system I know today. However, the best frame integrating video cameras may produce wonderful still frames when properly stacked and fully processed. You must determine what is appropriate for you.

Often described as permitting viewing deep space in real time (even in press reviews), I should point out that this is not quite an accurate description of the function of CCD video devices. No matter how impressive the results, what we see nevertheless is an accumulated simulation of a real-time view, albeit in a compounded relatively short period of time, and viewed on a monitor in a secondary sense at that. Therefore, to be accurate, these devices should still be regarded as imaging systems, rather than live "eyepieces to the stars." However, they certainly provide a rough parallel to the apparent aperture gain possible with an image intensifier, and in fact, sometimes go beyond; but remember, from the perspective of true live eyepiece viewing, some significant differences nevertheless remain. Cameras require monitor viewing, and produce less resolution than the most advanced intensifier tubes. Aside from the loss of instant connection between object and viewer, one needs to find, center, and focus the object in question, then manipulate the camera controls, before image building and viewing can begin. In the case of faint subjects, all of this may be challenging to do quickly or easily, but the viewing experience itself will still be very dramatic for most people. Considering the vast differences in cost, many people may be quite satisfied with the not-quite-live video substitute. It certainly is an impressive option, but do not think of it as the same thing. Even the frequency responses of some have certain things in common to image intensifiers, particularly in the red and infrared parts of the spectrum.

Some of the seemingly amazing images, sometimes shown off by proud owners of these CCD video systems, might lead one to believe that they can deliver more than they promise in the field. When seen at a small scale, many of these images may appear to be comparable in quality to those produced by the more expensive image intensifier, and the views it provides, live. Sometimes the pride is justified. In other cases, closer up, these video images often reveal flaws, both in definition and cleanness of image. Other images may look similar to relatively long time exposures, at least in brightness and impact, and you might easily be inclined to think these images always represent what would be seen at the time of viewing on the monitor. Again, sometimes they really are representative, of course, though in the case of many, what you may be seeing is the result of later image processing on a computer, in addition to careful frame selection and frame stacking of the CCD video system itself. Close up, some of these images exhibit all the "globs" and other artifacts that are characteristic of overprocessed CCD images. This does not negate the value of the results of the best examples, of course, but do not assume "live" views and processed still images are necessarily interchangeable. Many of the StellaCam images found within these pages presumably benefited from some later enhancement, in addition to normal frame integration. They are remarkable, nevertheless.

Adirondack Video Astronomy, probably the leading commercial authority and pioneer in video astronomy, has established a considerable reputation for introducing cutting-edge equipment into the marketplace. Their Astrovid StellaCam, its much more sensitive cousin, the StellaCam EX, and the soon to be released StellaCam II, are excellent examples of frame integrating CCD video camera systems. (The EX and new II versions offer hugely increased sensitivity for only a small amount more expense.) The StellaCam's comparatively low price (a quarter to a third of the Collins system) will make it a serious option for many people who want to experience much of what is written about within this book, but cannot justify or afford the significantly more expensive alternatives. Other less costly astronomical video cameras exist, but beware that most do not have frame-compounding capabilities, nor the important flexibility of manually adjustable gain. The true astronomical video pioneers, the Astrovid line probably represents the best of such products, though similar if not identical alternate systems to some of the StellaCam systems do exist elsewhere. However, my own good experience with Adirondack allows me to speak highly about the quality of product and service they offer. John E. Cordiale, of Adirondack, was particularly helpful in supplying information and images highly specific to this writing, regarding the StellaCam CCD video systems. His enthusiastic involvement speaks well for his company's products. (See Figure 2.5.)

Comprising two primary components, the StellaCam features a separate camera that attaches to the telescope focuser, and a control box, which is attached to the power supply by cable. You also will need to have a separate monitor. The system may therefore be operated with the control box situated directly next to the monitor and yet quite far from the telescope. The concept behind this camera is a somewhat different approach to previous CCD video cameras for deep space viewing and imaging. A monochrome system, it runs at the traditional video rate of 30 frames a second, but allows the user to stack (integrate) up to 128 frames at a time, and up to 256 frames in the new StellaCam II version. Another advantage

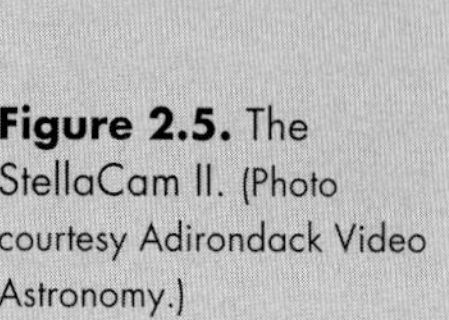

Figure 2.5. The StellaCam II. (Photo courtesy Adirondack Video Astronomy.)

of integrating frames is reduced noise and increased smoothness of the image. The StellaCam is quite impressive in this regard.

The best of what I have seen from the standard and EX versions is reminiscent of the gain provided by the standard Collins unit, though without quite the refined finesse of the I_3's live eyepiece image. The new StellaCam II goes further still, providing 18th magnitude reach(!) with an 8 inch telescope, or something like a five times gain in aperture, according to Adirondack. (This is closer, but still not equal to the effective apertures produced by the upgraded thin film tube of the I_3!) Promising to actually rival the performance of some cooled CCD cameras, it also has an extremely user-friendly control box, variable shutter speeds (an important feature), and three contrast (gamma) settings. The chip of any of the StellaCam units is quite sensitive to red and infrared light wavelengths, which also puts it somewhere in the direction at least of the spectral realm of a Generation III image intensifier. The image is displayed as it builds, and updated continuously, and the new StellaCam II version can even image very faint nebulae. Overall, StellaCam cameras, or other competing similar systems, represent a great advance, receiving nothing but critical praise from users worldwide.

A different, but still highly viable and high-quality alternative, an earlier but in some ways more sophisticated system (and definitely more costly), is the STV by SBIG (Santa Barbara Instrument Group). As far as I know, it is truly unique in the marketplace, and it first appeared in 1999. Despite the obvious similarities to the StellaCam, it approaches the subject with longer individual frame exposures, and is not based on a true moving video format. It does have, however, more processing features, as well as different ones. Proper use of either of these excellent video systems, however, enables the user to obtain quite good deep space images by real-time "stacking" of frames, and in the STV's case, also by building frame exposures over time (up to 10 minutes). The STV will cost you much more than the more recent integrating CCD video cameras, but does, of course, offer unique capabilities of its own, which I will detail shortly.

The STV is actually a complete, integrated monochrome system, with its own built-in monitor as an optional upgrade. Otherwise, viewing may be conducted on an external monitor. Consisting of a main tabletop unit with all of the controls (and monitor when so equipped), and a separate CCD video camera connected by cable to it, the STV offers remarkable capability. (Supplied in the configuration with internal monitor, its cost is actually closer to that of the Collins unit itself than the StellaCam, an important consideration.) Instead of being just a moving video system operating at 30 frames per second, it allows the user to expose up to approximately 2 frames per second, or dwell on any single frame for up to 10 minutes! It is also capable of stacking frames (the "track-and-accumulate mode"), as with the StellaCam. Extremely short individual frames are also possible, for imaging very bright subjects such as the moon and planets. However, the extended exposure feature is one of the STV's most significant and distinctive functions, as it gives the user an access to deep space quite unique in concept. For lengthy periods of time that need precise tracking, the STV also serves as an auto guider.

At the time of the STV observing session itself, the image displayed can be viewed continuously, even as it slowly builds in the track-and-accumulate mode. During this process, extended exposures may be gradually stacked on each other while the next is being made. At maximum image buildup possible, actual apparent magnitude may go significantly beyond the two to three times effective telescope aperture increase of other systems, even the image intensifier. However, because of the time involved, it is more closely akin to normal deep space imaging than live viewing, except that now we can see the results immediately following the observation. The STV also can provide more than one image scale, as well as being able to store a limited number of images. These can later be downloaded to computer. And in a final plus to its design, the system includes internal chip cooling. Pretty impressive, I think you'll agree. Certainly the STV is a device to consider seriously, if its imposing crossover design between live viewing and CCD imaging is something that holds particular appeal for you.

As with image intensifiers, electronic noise in video devices, though of different origins, may become noticeable at times. This is dependent on all the factors involved in the exposure of any image, and minimizing this undesirable effect is always something the user will wish to do. Certainly the compounding of frames minimizes this problem, in much the same way as recursive frame averagers (see below: "The Combinations of Both Approaches"). Because of the parallels in the applications of image intensifiers and video cameras, I should stress again that either form of viewing is actually quite in line with the purposes of this writing.

See the Appendix, "StellaCam II Highlights and Description"; also "CCD Video Cameras for Astronomy, and Accessories."

The Combinations of Both Approaches

The clarity and image refinement that can be seen in the direct eyepiece view through a good image intensifier is something not discussed very often, but I can assure you that it is amazing. In my opinion, it sets these devices apart from all other present systems, as it is not dependent on the lines of resolution utilized in video systems. However, some CCD video cameras can be connected readily to the Collins unit, should you want to share something of your observing sessions with others, either in simultaneous live viewing or by recording images. Adding a nonintegrating astronomical video camera (the Astrovid 2000) to my image intensifier is how all of my own deep space illustrations in this volume were taken. By this, I recorded real-time views as effectively and immediately as possible, while trying not to go beyond what can be seen, live, during later processing. There is no reason that other home-assembled intensifier devices could not also be coupled to a CCD video camera in the same way.

The single-frame images obtainable by combining image intensifiers and CCD video cameras, compared against those obtained in the frame integrated views of either of these new CCD video systems I have described, seem quite similar; they are just produced in a different way. However, either approach nevertheless results in a noticeable step lower in crispness compared to the live, intensified view through the telescope. It seems that all modern CCD video systems have a marked resemblance in concept. For example, the classic Astrovid 2000 CCD video system, only recently discontinued, is also typical of the general outline and concept of these, and looks much like all of the other Astrovid systems. On the surface at least, the main difference between StellaCam and the 2000, however, is the StellaCam's unique ability to compound frames. Adirondack also offer lenses for the StellaCam, not unlike that offered by Collins, for low-power, wide-angle viewing.

The I_3 may be connected directly to certain Astrovid video cameras for astronomical video, the connection being made with the addition of a special coupling/adapter, also manufactured by this company. Actually, the I_3 was originally configured to hook up to the Astrovid 2000 CCD video camera. This particular CCD video camera was introduced in 1998, and still seems an ideal camera, even though it is no longer available at this time. Check with Adirondack first, to connect the I_3 to their other newer systems. For home-built intensifier eyepiece systems, by placing the eyepiece optical component further from the fluorescent screen by means of an extension, it will also be possible to join the intensifier rig effectively to such a camera. You will need to experiment with this placement for the best results, but suffice it to say, an increased distance of 1 to 2 inches (25 to 50 mm) will be in the ballpark. The basic Astrovid design, consisting of a separate camera and control unit, still requires a separate monitor, of course. Still ideal to use in combination with the Collins unit, the Astrovid 2000 operates at the standard video frame rate of 30 frames per second, and remains a highly sensitive device suitable for registering faint astronomical subjects. It would still seem to me to provide easier control of images than virtually any other advanced CCD video camera, partly because it is so straightforward to use and sensitive in registration. It is logical that it should have been originally chosen for this particular application, and if you can obtain one in the secondhand marketplace for use with an image intensifier, you will not be disappointed. (See Figures 2.6 and 2.7.)

Utilizing the same Sony CCD chip as does the StellaCam, the Astrovid 2000 also offered a most direct control of the various optimal parameters for real-time astro-imaging. It is instantly possible to vary shutter speeds from 1/60 to 1/10,000 second with eight separate dial settings, manually affect the image contrast quickly during imaging with three manual gamma control settings (an important ingredient for deep space imaging), and have full manual adjustment over the

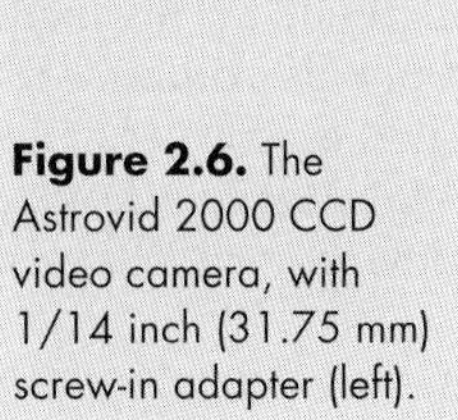

Figure 2.6. The Astrovid 2000 CCD video camera, with 1/14 inch (31.75 mm) screw-in adapter (left).

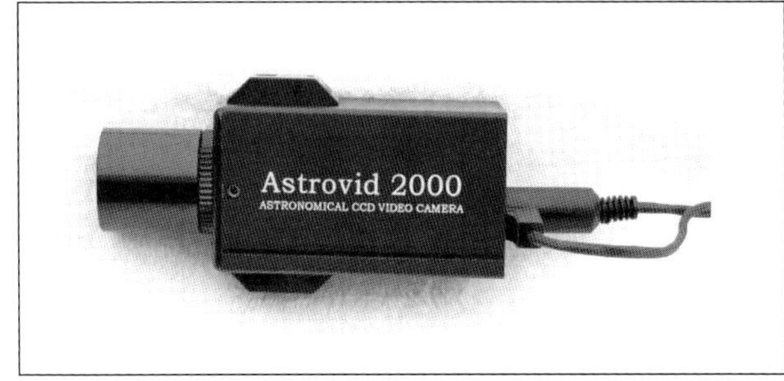

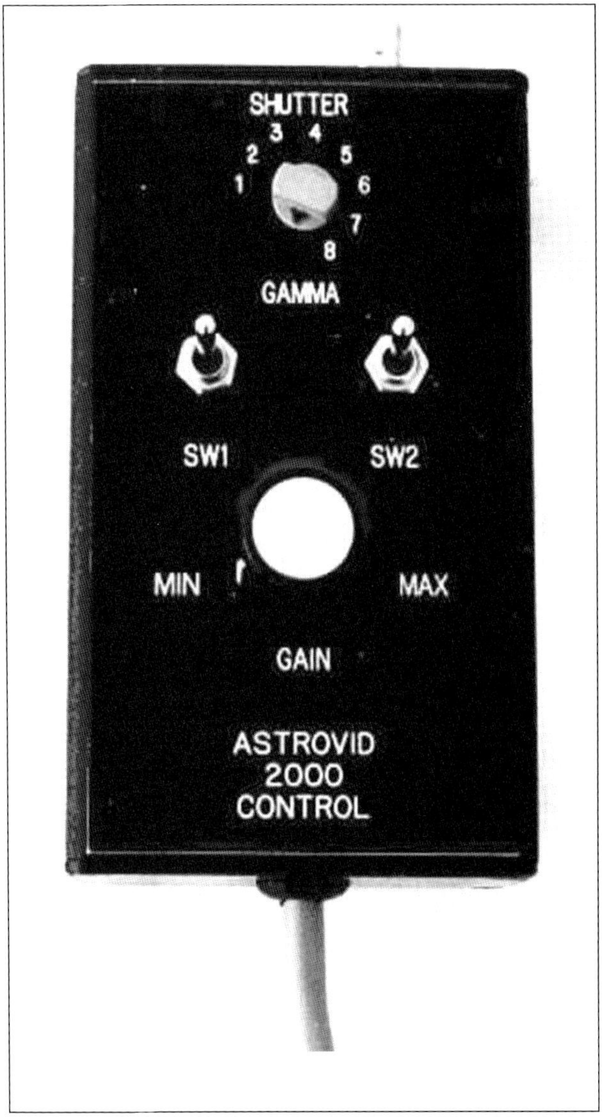

Figure 2.7. Astrovid 2000 control box.

gain with a continuously turning knob. (Aha! Analog controls; saints be praised!) The quality of this system together with the easy combinations of these various controls would seem to provide a ready flexibility that still puts it very high on the list for the applications I utilize personally. By this I refer to the adjustments that can be manually and quickly set, for the best image and least image noise. Without these and the simple accessibility they offer, many objects will be hard to register effectively, being either too bright (solar system), or too dim for optimum monitor viewing.

Connected to an image intensifier, such cameras are capable of imaging deep space objects in true real time, with a similar registration of magnitude as that of the eye itself during live intensified viewing. We should always bear in mind, though, that compared to any of the well-known time exposure imaging techniques, video is unlikely to equal them when recording real-time deep space images, no matter how we go about it. However, those deep space subjects that show themselves effectively in live viewing with our image intensifier will indeed register in a fairly representative manner with real-time video. This is the beauty of this concept; it is possible to capture the essence of what we have seen live, and create effective still imagery without processing, other than of the most basic kind.

The original video coupling and adapter offered by Collins were quite different from the coupling/adapter offered now, in that it reused the TeleVue eyepiece component in the optical path. Remarkably good results were possible using that particular configuration, but at best, it was still something of a compromise in optical design, and was only able to utilize the central portion of the field. To do otherwise apparently would have resulted in star images near the edges being highly distorted. Additionally, it provided no adjustable iris of its own; when using an image intensifier in video applications, we no longer have the uniquely sensitive eye as the sole means of image registration. The lack of such an adjustable iris in the old unit meant that these differences in sensitivity could not be compensated for when imaging faint objects; details could not be enhanced for brightness in the primary image before they reached the camera. The old video adapter system has now been replaced by the present superior single unit, which has its own integral optics with adjustable iris. It is also available to overseas customers. You may be able to design a similar unit yourself for a home-built system by adapting an old camera or video lens. Unfortunately, I cannot provide details of how to do this, and some experimentation would seem to be at least partly in order. (See Figure 2.8.)

This new adapter produces good definition to near the edge of the field, the size of which has been more than doubled. Along with the adjustable iris also comes a flexibility of adjustment of the image brightness, and noticeably increased image quality, and resolution provided by its better-matched optics. The image intensifier part of the I_3 unit is simply detached from its TeleVue eyepiece component, and threaded into the video adapter, which is in turn threaded into the CCD video camera. A very stable combined unit results, of about 10 inches in length, shorter and altogether less prone to optical misalignments than the old one. I have found it to be infinitely easier to produce the most advantageous brightness and contrast for each subject, by working with both the adjustable iris of the video adapter and the gain and shutter speed settings on the camera control. I

Figure 2.8. The new Collins Electro Optics video adapter.

was restricted with the old system to fewer options, although it produced quite acceptable results most of the time. Indeed, I imaged virtually all of the deep space illustrations in my previous book with that old system.

You will need to be careful setting the iris, however, since brighter stars may appear to be somewhat triangular if it is opened too wide, as well as rendering relatively luminous parts of the subject out of focus. Sometimes, however, because we are dealing with real-time imaging, I have found a slight amount of the triangular star syndrome to be an acceptable compromise. This arises when I am trying to obtain a decent registration of faint subjects or details if the subject's relative magnitude is much dimmer than surrounding stars, although I endeavor to keep the effect to a minimum. (See Figure 2.9.)

Figure 2.9. The Collins I_3 with the new video adapter, attached to my Astrovid 2000 CCD video camera and JMI 18 inch telescope at desert location.

You will also be aware of the effect of coma in many of these images, a characteristic of fast Newtonians and not the adapter; stars at the edge of the image sometimes seem to trail off like comets. This is because we no longer have the corrective action of advanced eyepieces such as those offered by TeleVue, and it is certainly not due to the intensifier's optical system itself. For the most part, the low magnifications employed with deep space make this only a marginal issue, but with higher powers it does become more noticeable. For this type of viewing or imaging, it seems to me that such issues are hardly worth being concerned with, but if it really bothers you, by all means use a coma corrector, such as TeleVue's Paracorr. These optical correctors are simple, and do the job across most of the field.

The video adapter system I use produces an image scale slightly less than half of what was obtainable with the earlier one, and the round edges of the I_3's field can be seen in the corners of the monitor. Far from being a disadvantage, what this actually means is that we finally have the capability to see the full width of the intensifier's phosphor screen itself, since definition has been so greatly improved to the edges. It also illustrates that previously we were only able to utilize just under half of that total field, and hence had a coarser image resolution than before. In scale, the image you will see on the screen is something more akin to the impression of the live view obtainable with a low-power eyepiece in any given telescope. This is just a guide, of course, and based only on my own reaction to what I see. Collins Electro Optics provides the following formula for calculating field size of the image with the new adapter system: 2369/focal length of telescope (in inches). In my case, that corresponds to 2369/81 = 29.25 arc minutes for my 18 inch reflector. (To adapt this formula from focal lengths in metric dimensions, divide total millimeters by 25.4.)

Low subject brightness is much easier to deal with by selecting the reduced scale. If you want to know the image size with any greater magnification, simply adjust the figure produced by the above method according to the power of the Barlow lens used in conjunction with the system. With a 2X Barlow lens, for example, the image scale can be brought back up to something close to that of the old adapter system, although the round edges of the field are still present, of course. This is because the phosphor screen size we are amplifying does not change; remember, because the effect of the Barlow occurs before light enters the intensifier, the appearance of the screen outline has nothing to do with image size. None of this ultimately matters when it comes to extracting still images; if you want to outline them in rectangular or square boundaries it is easy enough to set them any way you wish. Be aware, nevertheless, of these quirks with the new adapter, so you will not be surprised. Bear in mind also that the image that you see on the monitor may not be exactly centered with regard to the actual image being recorded, due to slight misalignments of the components relative to the camera's CCD chip, the length of the optical train, small inaccuracies within this optical train, and also the monitor's centering adjustment. Because of the latter factor, you will find that any irregularities in this regard are substantially corrected when you examine the images after they have been downloaded to computer. Therefore, remember to center your images in the middle of the true field, and not as they appear on your monitor; this is not difficult if you center according to the field's partially visible circular boundary.

There are obvious additional advantages with the new coupling system: with the lower effective power than possible in the past, it is easy to switch back and forth between the much wider field of view it allows, or larger scales, by adding different Barlows. Reverting to the old adapter system in combination with a 2X Barlow can more than double the power again, but most subjects are too faint for this to be effective. For those relative few that can handle the additional power, it would be better to invest in a more powerful Barlow, such as TeleVue's 5X Powermate, and simply apply it to the new video adapter system instead. I now routinely use this same 5X Barlow for great image scale and the high-grade optics it provides for my planetary and lunar imaging. Such powerful Barlows, which are 1-1/4 inch diameter (31.75 mm), are just fine in the wide optical paths some amateurs' larger aperture telescopes are likely to utilize; by default, only a small portion of the field is used at these high powers. Therefore there is really no reason to manufacture what would be much more expensive 2 inch (51 mm) Barlows for ultrahigh powers. The combination should produce roughly similar image scale as the maximum obtainable before with the old system, at least for those deep space objects bright enough to benefit. Deep space objects that respond well to extra high powers usually are relatively well defined, fairly bright, and compact. Most likely, this means globular star clusters and planetaries. On many of these subjects, a relatively high power will actually be found to be more or less essential, since the objects often may be either small and intense (planetaries), or compact, and in need of "prying apart" (globular clusters). With all of these possibilities, we now have the best of all worlds and capabilities. So, all in all, for intensified video applications, I think all that I am likely to need is covered, but I am hanging onto the old coupling and adapter for possible special uses, just in case. I did, indeed, use the old system for just a few of the images in this book. The view on the monitor, totally filling the corners, is one such reason occasionally to put it back to work, but generally speaking, the new adapter has made my old system obsolete.

The resolution of detail will also be affected by matching as closely as possible the ideal image scale for your own telescope and video imaging system. This is dependent on the number of pixels of the chip itself, and the telescope's optical specifics, assuming the brightness of the subject is sufficient to allow any range of power. Usually, only the lowest power will be found to be adequately illuminated, and so the theoretical ideals hardly apply. Where we do have great subject brightness, we can try to take advantage of registering an ideal proportion of pixels to the potential resolution of the whole setup, but take it from me, this is likely a moot point most of the time.

There is one important other wrinkle when it comes to video applications with image intensifiers in particular. Unfortunately, it also involves cost! If we want to have smooth intensified video on monitor or usable recorded clips, or even to be able to freeze acceptable single frames later on computer without stacking and refining, we will need an additional piece of equipment in the chain. This is because the electronic scintillation and noise produced by the intensifier becomes an issue in video, and all the more so as we adjust gain and contrast to register our intensified subjects effectively. This downside of adjusting these parameters will render any imaging substantially less effective than is the live view through the eyepiece component, and all the more when we select single frames later. Unfortunately, such frames become virtually useless, and the differences from one to another can only defy description. You will wonder if the individual frames are of the same subject! Luckily there is a solution. We will need a recursive frame averager for use with our image intensifier (Figure 2.10); in this matter we have no choice if we want to have the smoothest intensified video, and usable stills. These accessories, of which a fine example is also available from Collins Electro Optics, takes multiple video frames (up to 16), and removes most electronic artifacts not common to all by averaging them. See the Appendix for some other companies supplying recursive frame averagers.

The result of incorporating a frame averager into the system is a near-perfect smoothing out of either moving or still images, making them suitable for all of

Figure 2.10. Collins Electro Optics Recursive Frame Averager.

our purposes. The application approximates what the eye sees in live viewing, except now the image is on a screen. Without such equipment, all of our extracted intensified frames, sadly, will be near worthless.

In what may seem to be another tantalizing combination of camera and intensifier, it is possible to connect the StellaCam with the I_3 with the same couplings as before. This would seem to offer the best of both worlds, on the surface at least, and should be theoretically possible with virtually any other system as well. Combining this type of camera with an image intensifier would certainly be an acceptable approach for general video hookup applications, and once was suggested by Adirondack. The STV may also be connected in this manner to expand the capabilities of video. However, both viewing approaches (compounding video systems, or intensifiers) were designed as complete deep space viewing entities in themselves, but let us examine the possibility of combining their unique attributes in joint usage at least in theory. The StellaCam would seem to offer similar single-frame capabilities as the Astrovid 2000 when combined with an intensifier, since it utilizes the same CCD chip; the EX version would seem all the more so since it uses an even more light-sensitive chip. I cannot claim to know how close they come in a direct frame-versus-frame comparison. However, this aside, the easy analog controls and manual control of contrast of the Astrovid 2000 still remain among its distinct advantages, even at this level.

More to the point, however, is how frame compounding works in conjunction with the intensifier. Electronic noise of the image intensifier in video applications can be considerable. Unfortunately, when the original StellaCam is used to compound intensified frames, the camera appears to preserve one unaveraged frame on top of all of those that have been stacked. This results in a far less than ideal image. So it would seem to me that there is a real limit on the combined use of the special attributes of these two particular devices. Adirondack no longer appears to recommend this combination. Using the STV with an intensifier may be a better proposition, but I will leave any of these possibilities for your own experimentation and curiosity.

If you are interested in extracting the maximum that is contained within the images taken from your own live video, multiple frame selection, stacking, and processing on your computer can produce results that in some instances do indeed begin to approach time exposures of deep space subjects. There are many imaging software applications, such as Adobe Photoshop, which will enable you to process your selected still images to an even greater degree than is presented with the images in this volume, should you wish. (By the way, fine planetary color images are also possible with CCD video cameras, by exposing several images through different color filters, just as in standard CCD imaging. Proper processing of these has produced fine images as good as any CCD image I have seen. Astrovid also offers the PlanetCam for instant color planetary imaging, but the results fall short of what is possible with monochrome cameras and triple filter exposures.) For any such advanced uses, it will be necessary to utilize many of the same computer programs that CCD and certain photographic applications rely upon. These refine and enhance the images considerably more than I wished to do for this writing, to a degree that far more detail will emerge than can be seen live. However, this is not my primary direction. My purpose in this book is to give a reasonable expectation of what you can expect to see, live, and so my own imaging methods tend toward the same philosophy: that of minimal processing, and simplicity. If you wish to explore further, I would strongly recommend the book *Video Astronomy* by Massey, Douglas, and Dobbins, (Sky Publishing). You will find all kinds of information on extracting the maximum from your video images, and specifics of video astronomical applications in general. All of the above fall outside the scope of this book.

Since video images can be recorded onto any video recorder, quality and type of recorder are obviously important factors. Of all of them, digital VCRs will essentially preserve all there is in the image without degradation. However, if these images are analog and we wish to view or store any of them on computer, it will be necessary to feed them later to the hard drive in digital signals. You can accomplish this with a Sony Media Converter, which takes analog signals from the analog video and converts them to digital signals for interfacing directly to computer. In my own case, the interfacing works via "Firewire" to one of my iMacs, viewed and stored though simple iMovie software. Individual still images can later be extracted and further mildly enhanced with this software to more closely approximate the view through the eyepiece, live.

As a slightly different concept, and not too far removed from real-time imaging, regular still digital cameras also can be coupled to image intensifiers to provide spectacular results in only a few seconds of exposure. The resolution and brilliance obtainable is beyond that of any of the video systems, and also frequently beyond that attainable by live viewing. Again, since this type of imaging is beyond the scope of this book, I will leave it to your own exploration.

In conclusion, I should stress again that while the results, utilizing the means outlined, can be very good indeed, the firsthand view provided by the eye (as opposed to the screen) is still the most sensitive tool for the registration of images. You already know the only choice of observing this particular approach is limited to. Fear not; you will still enjoy video images immensely, obtained by any of the methods I describe. You can be sure that more methods will eventually be on the way.

Eyepiece or Video – Which Path Will You Take?

We should take a moment to consider the likely impact that video applications, such as those we have been discussing, ultimately will have on amateur astronomy. Professional astronomers have been utilizing "remote" astronomy (albeit, not exactly video!) for years; few of them actually spend time at the eyepiece of the great telescopes they utilize. Many of these same advances are slowly becoming available to the amateur, including even an elementary form of adaptive optics (as introduced by SBIG a few years ago) for CCD imaging. These new technologies would seem to point to two distinct amateurs' camps. One will be centered around the adherence to more traditional visual astronomy, hopefully taking advantage of technical advances to enhance that ideal. The other will produce a type of observer not concerned at all with being connected to the real-time view. I believe I fit snugly into the first category, although I will confess that even this is already somewhat changed from its past (more in Chapter 7).

The closer one comes to simply viewing an image directly through the telescope, the better one will appreciate the special experience that live viewing offers. I believe this finally may have led us to the truth concerning the lack of acceptance of image intensifiers in the amateur community (which does not yet seem to have translated to astronomical video cameras). Coupled with their expense, image intensifiers are seen as falling into neither camp: live viewing, or video! After all, the image is not unlike video, at least in that it is projected onto a small internal screen; but unlike video, it is integral with the telescope and not processed in any way, being viewed in the same manner as a conventional image. However, take it from me, the practical effect of using one is virtually the same as with any standard eyepiece. It falls clearly into real-time, live viewing. As amateurs, most of us must realize that our primary role is something less than the scientific ideal; most of us are more like sightseers in the universe. I think that our role is more clearly realized through the live, visual ideal, although I will concede that truly astounding photographic results are being obtained by those willing to invest the time and passion to pursue CCD imaging techniques. Feel free to disagree on your role, though you may not convince me to change mine!

If the whim strikes you, another unique value of the image intensifier for live viewing or imaging is the study of novae and meteors. In *Visual Astronomy in the Suburbs*, I wrote of stumbling on a supernova in NGC 3190 while making routine observations at my suburban home, and without the slightest difficulty. The easy visibility of this object was remarkable, and there was no straining to see it, even initially; the identity of the stellar explosion left little doubt in the mind or eye. Even an inexperienced observer would have had no difficulty seeing it. It also showed up prominently on the simple video images I included. Confirmation of what I had seen was found by accessing the International Supernovae Network (see the Appendix). Other observers have found considerable benefit with image intensifiers in meteor observing. Certainly, hardly a night goes by without satellites and meteors crossing the intensified field of view with surprising regularity. In the case of ever-present satellites and space junk, which become frequently visible this way, it is sobering to consider just how many things are up there!

Of course, most solar system observing (e.g., the moon or bright planets) with an image intensifier is out of the question, not only because of the lack of need or advantage, but especially because of the damage that would result to the sensitive phosphor screen. However, in cases of extreme faintness, such as the far outer planets, comets and asteroids, an image intensifier does provide invaluable service. Neptune is a questionable target. In larger amateur scopes, it may be too bright. I have yet to train my intensifier on it for fear of damage. Uranus is worth a try, and may even show us banding, if we can only obtain sufficient magnification or image scale; a powerful Barlow lens may be necessary. Pluto becomes readily discernible in much smaller instruments than would otherwise be the case. Regardless, there should be no need to struggle to see it, assuming one can know its exact location. Its magnitude should be well within the intensifier's grasp for almost any telescope other than the truly small. When in doubt, it is best to wait until the planet is in the vicinity of a relatively bright star, plotted on a map, as periodically described in the monthly astronomical magazines. Frame integrating video cameras can also do valuable service in all of this, however, as there is no danger of "blowing their tubes" with nearby bright objects!

All of my own deep space images illustrating the upcoming chapters were digitally recorded in conjunction with my 18 inch reflector. They all utilized the I_3 coupled to the now-legendary Astrovid 2000 CCD video camera. The particular usage I get from this camera will probably keep it as my first choice for some time yet. The images presented here are all single frames, presented with minimal processing; they have not been compounded or superprocessed in any way. Usually, some simple adjustments in the computer were made in order to give a decent approximation of the live view, amounting only to minor adjustments of contrast and brightness. Because you will notice, at the time of observation, that images leave an impression noticeably more brilliant than the way they appear later to you as recorded images, this type of adjustment better simulates the eye's and mind's actual registration of the live images through the intensifier (or even the views on a monitor via any of the systems I have described). No, you have not lost your mind! Some of the reason for this is the effect of what amounts to a partial attainment of dark adaptation at the time of viewing itself. Image intensifiers and cameras may also differ in their specific registration from that of the eye. Even allowing for all of this, together with the type of adjustments that are easily performed later to the contrast and brightness, there still is no way to recreate in a printed image the luminescence the eye itself perceives at the time of viewing, either directly or on a monitor. At times, these adjustments to the still frames result in a certain amount of pixillation of the image, especially in the texture of the background sky. Considering what can be seen on the printed page using this simple method, the results, to me, seem astounding nevertheless. Hopefully you will forgive any of the various slight negative effects as I endeavor to provide a reasonable impression of the way the subject itself appeared at the time of viewing.

We now need to turn our attention to the other important factor in our quest to see ever more: dark sky sites.

CHAPTER THREE

The Dark Sky Site

Visual Astronomy in the Suburbs dealt with the various approaches we might take when pursuing the goal of good viewing from suburban locations, where the challenges can be considerable. As a resident of southern California, I am routinely subjected to some of the most polluted air (largely light and haze) that an amateur observer is likely to encounter. However, I feel truly spoiled by some of the extraordinary dark sky sites that we also have within relatively easy reach. You would think that with such massive light polluters as the nearly continuous 150 mile (241 km) sprawl of Los Angeles, as well as nearby San Diego, truly dark or transparent sky would never be attainable anywhere in this part of the world. Ironically, some of the sites readily accessible to both of these centers have been ranked among the best anywhere in the United States! Most are around two to three hours' drive time away from my home. I have listed in the Appendix (under "Resources") Web sites you might investigate in searching for dark sky sites accessible from your own location.

So, what qualifies as a dark sky site? The best locations will be situated, on average, 100 to 150 miles from any urban center. Fulfilling that requirement may prove to be a tall order in this day and age, but in North America there are still such locations in plenty, especially west of the Great Divide. The very darkest of them are well inland, but even in southern California, we can still come close to the ideal for the near attainment of darkness. It will be harder to find such suitable sites in many European countries, because of the greater proximities of so many urban centers to each other, so there will be fewer options as you assess your viewing potential. However, their general light pollution is likely to be somewhat less severe than in the United States. If you cannot easily access anywhere that would qualify as remote enough, it is still possible to find places to observe that provide near-dark skies; some of these may be surprisingly close to urban centers. Other factors, such as altitude, placements of mountain ranges, coastal conditions, humidity, and so on, all play an important role in the overall effectiveness of any given site, and so it is not as simple as just the physical distance they lie from cities. Even within the greater Los Angeles area, home to some of the worst light pollution known to humanity, there are some remarkably good semi-remote sites, well sheltered by low coastal mountains and canyons.

Our own geographic position also plays a role in how much and what portion of the celestial sphere we can see from any given location. Unfortunately for many of us in the Northern Hemisphere, some of the finest sights in the visible universe happen to lie in the Southern Hemisphere! Luckily, many of these southern wonders are also accessible from some of the more temperate Northern Hemisphere locations, and we can take advantage of our local geography in order to maximize the potential. This will often mean visiting not only the most geographically southern places we can reach, but the higher local altitudes wherever possible. Because of its position on the globe and natural attributes, southern California offers pretty good prospects in this regard. There are some wonderful high desert locations and good mountain ranges within easy reach of the population centers, which have both good air conditions and moderate to considerable altitudes. From such elevations we have not only our best southerly horizon, but also the least thickness of atmosphere to contend with when viewing Southern Hemisphere objects lying close to the horizon.

Better still, geographically at least, are more southerly locations such as Hawaii, south Florida, and Mexico. Yet, they are still in the Northern Hemisphere. In many ways they have some of the best viewing potential anywhere because they have access to almost all of the great sights of both hemispheres. The downside is the frequent humidity in these locations with the consequent likely decreased light transmission, a bane of deep space observing. Sadly, from many of the Northern Hemisphere locations at latitudes much higher than the 34° of southern California, observers will be unable to access most of the grandest Southern Hemisphere sights, as they will always lie below the horizon. Perhaps these observers will one day have the chance to experience them from other places.

For the images used in this book, I selected one of the most southerly places, as well as the darkest, to which I could easily transport my equipment. There are a host of potential dark sky sites I could have visited, but Mount Laguna and Anza-Borrego State Park are about as far south as you can get in this region, also being blessed with favorable conditions for most of their available campsites. These large state parks are located about 50-100 miles (80-160 km) east of San Diego, and lie at 2,500-6,000 feet (762-1830 m) on average. Although they are not so remote as to put them far out of easy reach, they usually successfully escape most urban skyglow, particularly when my old bugaboo, a blanket of marine layer, fairly common at my home site, comes a few miles inland and covers up the city lights westward near the coast. The high desert sites have another special advantage over most others: the frequent dry air is excellent for a wide spectrum of light transmission, an important ingredient for deep space and enhanced live viewing. This attribute has led many experienced observers to rate them highest among viewing sites in the United States. (See Figure 3.1.)

Although the following comments on the locations I selected will be of most interest to observers residing in southern California, I thought, in describing a

Figure 3.1. Blair Valley site at Anza-Borrego State Park.

little of the area and circumstances that were so pivotal in the making of this book, it might also be of interest and help to others. In addition, it will help to complete the mental image you formulate, and maybe pass on some of my own enthusiasm for setting up at remote locations.

I had heard that the Meadow Loop campsite at Mount Laguna was exceptional, perhaps even the best site of all in the region. Thus, originally it was my prime pick. However, although this site, which is contained within a state park, offers many other campsites within its perimeter, I soon found out that booking ahead is the name of the game (see the Appendix). With weather never being a sure thing for astronomers, it is not always practical to predict far in advance when we can actually go out on location. Usually, by the time I knew with reasonable certainty what conditions were in my future, I found myself striking out when it came to site availability at Mount Laguna. Faced with such scheduling problems, I soon found that many sites at another state park, Anza-Borrego Desert State Park (a little further northwest), required no advance booking. (See Figure 3.2.)

The open desert sites in this vast park are only marginally improved as camp facilities, which certainly has a negative impact. However, the positive attribute is that not only are they more likely to be available, but they are also typically free from nearby camping neighbors with attendant lights, or the possibility of pestering and unwelcome nocturnal visitors (at least, of the human variety!); more later. This is a more important factor to consider in your own planning, whichever dark sky site you decide to visit.

Something else that Anza-Borrego offers over most other state parks is that one is allowed to set up camp virtually anywhere within its perimeter. This certainly provides a workable alternative if the official sites are all too busy, and there are many opportunities for improvising places to set up for a night of viewing. At

Figure 3.2. The JMI NGT-18, after preparation for a night of observing.

their best, the skies at higher elevations anywhere at this desert location are also quite outstanding. If not quite possessing the rarified air of Mount Laguna, they nevertheless feature wonderfully transparent nights, glowing with cosmic light. On good nights at Anza-Borrego I have seldom seen more stunning skies; most of the images in this book were obtained there. Should you wish to visit Anza-Borrego, I would say that the best conditions I have found overall are at Blair Valley, on the western side of the park. It should come as no surprise that this site was to become my preference for setting up for a night of viewing or imaging. The remoteness is only underscored at nightfall by the howling packs of coyotes all around; spooky to be sure, but I am assured they seldom pose any threat to humans, or even a subhuman species known as amateur astronomers! These critters, supposedly nocturnal animals (as well as desert foxes, beautiful creatures to be sure), sometimes ventured almost next to me where I set up, and the boldest coyotes did so even during daylight hours (Figure 3.3). Rattlesnakes (not so beautiful) are quite common during warmer times; I understand cougars are not unknown! A park ranger warned me that astronomers and others who stay up at night should beware, and that we should always view in "packs" of at least two people (not that rangers had actually heard of an attack on our own particular astronomer species yet!).

The coyotes I have seen at Anza-Borrego are unusually beautiful and groomed, considering the usual lot of these animals, and they seem quite healthy and well fed, instead of the more familiar straggly varieties. They appear to be either totally dispassionate, or sometimes just curious; they are known to associate human presence with food scraps. After nightfall, a flashlight beam sometimes reveals numerous pairs of eyes around the site, starlit points to be sure, but not quite the starlit points we had in mind! Other times, all you will see are bushy

Figure 3.3. A beautiful visiting coyote at desert site. (Photo courtesy Andrew Shulman.)

tails scurrying off into the distance. To date, I have not seen them behaving aggressively, although the scrupulous cleanup of food scraps is important in order to keep them that way. All things being equal, I have not yet plucked up the courage to venture into the wilderness alone for a night under the stars. Not wishing to become part of the food chain, I think this is wise, if maybe a little overcautious.

Special Needs Away from Home

From the practical standpoint, there are a number of special considerations we have to deal with when we move our astronomical gear, and especially when this is to a remote site. When using the telescope in combination with enhancing devices it is more complex still, especially when we wish to record images or to provide monitor viewing. For visual sessions with image intensifier only, this is a much simpler issue than with any other system.

Perhaps the first question to ponder is the need for electric power to run our various devices. If we are to provide power for a monitor viewing or imaging system in the same manner as at our home base, the camera, the recursive frame averager, the analog-to-digital converter, and the recording device, be it a computer or video recorder, all have their own specific power needs. Often, differing voltages and special electrical considerations of each piece of gear present more problems than may seem practical to take on at a remote site. Even having the necessary battery packs and all the needed output sockets at our disposal, or even having the use of an independent power supply (some official astronomical and camping sites provide this), only partly resolves the issue. It is challenging enough having configured the power supplies to everything, without having to set up all of this equipment in an unfamiliar environment, coordinate its use, and still avoid tripping on all the wires in the dark! On this last point, at my suburban location, the limited darkness available always meant I could vaguely see my way around the telescope and attached equipment, at least to some degree. (This still wasn't always enough to prevent me stumbling or dropping one of my prize eyepieces, twice, onto concrete.) If you become used to this mode of operation in the suburbs, it will come as a great shock to suddenly find yourself immersed in the true inky blackness of a moonless night at a true dark sky site.

For the simplest possible setup on location, any of the various video devices I describe require 12 volts, and could actually be powered during taping by plugging them into my telescope's own internal power supply, also 12 volts. Your telescope may also have such a potential power source. As for my Astrovid 2000, the camera's manufacturer, Adirondack Video Astronomy, states that as long as a 500 mA fuse is wired in series with the positive lead (going to the camera), it is safe to do this. Luckily, the NGT-18 has two power outlets, so if I choose to use its internal power only as my complete power source, it is simple enough. In this case, I take care to unplug the camera between the times I am recording in order to save power. I do not, however, recommend this approach as the best way to proceed.

A far better and more reliable way to be sure of sufficient power throughout the session at a remote site, not only for the telescope but any associated video devices, is to have the use of an external power pack. Kendrick Astro Instruments supply several varieties of these, and they have large enough capacities to take care of any task in hand. The one I personally use features two individual 12 volt power outputs. Other versions feature a single output, as many as four outputs, or can provide 18 volts, needed by some telescopes. The units are of high quality, rugged, very well made, of large capacity, and designed with astronomy specifically in mind. I strongly recommend such a battery pack, regardless of your telescope's own power capability. You do not want to "run out of gas" in the middle of the night, especially after you've gone to all the trouble of lugging your equipment to a faraway site. (See Figures 3.4a and b.)

Substantially less expensive rechargeable units, though of a lesser electrical capacity, are marketed by a number of companies, including Celestron and Orion. Basically the same unit flying under different flags, they feature twin double 12 volt outlets, with three additional outlets of 3, 6, and 9 volts, a steady flashing red or white light, plus a truly awesome main beam. (It is more like a searchlight!) This unit will probably supply sufficient power for most people's total needs, but for me, it was the mighty beam of its main light that made it a prerequisite at remote sites. For me, the access to such luxury of brilliant and sustainable illumination in the wild made it the best bargain of all at less than $60 US (£33 UK). At this price, it does not even have to hold up over the long term. Aside

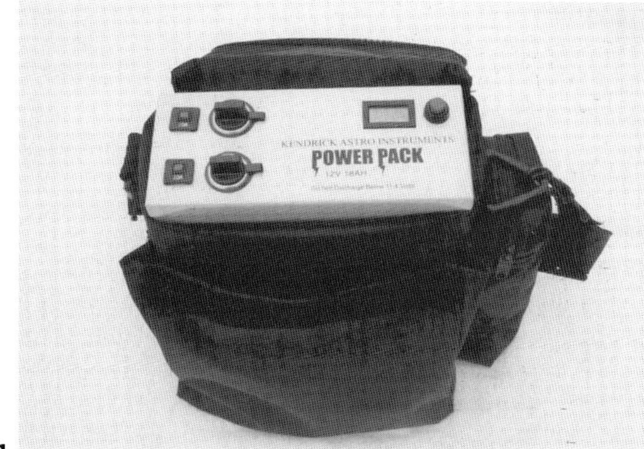

a

b

Figure 3.4. **a.** Closeup: Kendrick double outlet rechargeable power pack. **b.** Kendrick power pack set up on location by telescope base.

from the backup capability it provides, having the potential for powering additional devices of differing voltages makes it a tremendously useful accessory.

When it comes to imaging sessions in remote locations, I have always favored the simplest approach possible. The type of imaging I engage in is only a means of sharing something of the live experience with others. Extracting the maximum possible from images, live, or time exposed made by sophisticated techniques, is not my primary objective. The reason I take this approach is directly in line with the preservation of the "live" ideal, my basic mindset. At this time, only CCD video imaging allows me to do this, a true live imaging system without the element of time exposure. I therefore decided to use essentially this, the same overall approach for recording the video images that I had utilized in *Visual Astronomy in the Suburbs*. This consisted of telescope, image intensifier, CCD video camera feeding a stream of images into frame averager, then converted to digital feed for final inputting to computer hard drive. However, being far from home, I decided to simplify things and turn the process essentially into a two-stage affair instead. This involved a slightly different means and sequence of events: recording the basic, raw moving video images, and leaving frame averaging, computer processing, single frame image selection and storage until my return home. Regardless of the imaging method we employ, it is always important to keep simplicity in mind if we do not want technology bugs biting us during our special observing outings.

Although virtually any camcorder will suffice, the digital variety in the form of a portable VCR, with a relatively large built-in viewing monitor and its own battery power, will do better. The high-quality images obtainable by digital technology means there is no longer much of an advantage to feeding images directly onto computer hard drive, as I had always done in the past. One company, apparently only Sony, makes a rechargeable internally powered portable digital video recorder with built-in screen, the Sony GVD-800 Video Walkman (Figure 3.5). This provided the solution I was looking for, although priced in the region of $1,000 (£620 UK) it does not come cheaply. The recorder comes with a 4 inch (101 mm) built-in color monitor (though the images from the camera are black-and-white), so it is easy to see what you are taping. The advantage of this becomes very significant when you are trying to achieve good focus. The video recorder also serves to keep my computer hard drive uncluttered, because of the high memory requirements for video clips recorded on computer hard drive. This simple equipment lineup has become my method of choice for image storage now, even when imaging at my home location. Fortunately, this unique little

Figure 3.5. Sony GVD-800 Video Walkman.

machine also allows the input of analog video signals, so there is no need to add any more equipment to the chain when in the field. It also exports video in either analog or digital formats, so it is entirely compatible with all of our needs.

Complicating matters in the field, it always seems to happen that glitches of one kind or another seem to occur at inopportune times, and especially when we are powerless to do anything about them. Murphy's Law is apparently an invariable at inconvenient places. In the very first imaging session for this book, having just driven 110 miles (177 km) and set up all my gear, I found that the declination drive motor of my telescope was not working. Not being able to solve the problem at the time, I limped through the night, grappling and struggling with the telescope the entire time, only to find upon my return home that just a tiny piece of desert grit had lodged into a connecting plug. This had caused all the trouble.

On my second trip, the static electricity of the dry desert air caused a discharge to the telescope's digital circles, and also the drive module, any time I touched the control. This shut down the digital circles' module and disabled the drive several times. Only after I had figured out the problem did I take care to discharge my hands before touching any equipment. On the same outing, and almost from the start, the CCD video camera's gain control ceased working, stuck in maximum position. This was not too serious for deep space, where it would be maxed out most of the time anyway, although it was at times problematic. Later, a wire to the camera broke and blew the fuse. I did have one spare fuse, but because of my inability on location to make a permanent wire repair, it blew again an hour or so later, and finally shut down my session. Luckily, I had made it through most of the night.

On a later trip, the temperature plunged way below that which had been predicted (to well below freezing), and resulted in heavy condensation forming on the secondary mirror, ultimately freezing over! The telescope's cloth light shroud was also covered in ice, as was most of the telescope and the attendant observers! The only solution was to disassemble the upper portion of the telescope, take it to the van, and with the engine and heater running for an hour, thaw it out (along with the observers)! Another time, the weather forecast (yes, that again) was wrong in every conceivable way; everything from high winds to dust, haze, high cloud, and even rain clouds conspired to ruin the session. It could hardly have been worse, the strong winds making even boiling water for a hot drink a challenge. On this occasion, I made no attempt even to unload my equipment, eventually turning around reluctantly for home – three hundred miles of driving and all the preparation it entailed for nothing. Even our attempt to heat some food ended up with it blowing off the picnic tables into the desert sand!

The moral of these stories is that even simple plans, assemblies, or anything else can betray us when we are in unfamiliar or hostile surroundings. Indeed, because of our friend Mr. Murphy, unforeseen disasters invariably seem to single out these special sessions to cause us the most grief! Therefore, plan for it! Reduce the chances of technical malfunctions, at least, by reducing your gear, and also by thinking ahead of other potential problems. Maybe you can circumvent them by a taking different course of action when you are away from home. Also, use heavy-duty components that do not fall apart or break at the slightest bump; in the wild, it is not always easy to be graceful and completely choreographed like a ballet dancer. I can live with the minimum of these kinds of distraction and headaches in the boondocks, where such nuisances are greatly magnified, instead of the objects we wish to see.

Before We Pack!

Perhaps the greatest potential hazard to get in the way of a successful night's viewing is the weather itself. Historically, it was always difficult for the amateur observer to predict with any degree of certainty the weather conditions for an upcoming dark sky trip. Nothing was worse (or still is for that matter), than going to all the trouble of hauling all of our gear to a remote site, only to find that the weather lets us down! (See Figure 3.6.) I suspect my own experiences are typical. Fortunately, today there are some excellent weather resources for us to keep track of what is going on in the night skies. It is possible to have an accurate picture at least most of the time; there is still no accounting for the exceptions(!). I have been usually quite successful with the Web site www.weather.com for some time, and indeed still take advantage of it for long-range planning. A division of The Weather Channel, it features comprehensive reports worldwide. Its 10-day forecast is still invaluable and more reliable than most sources, particularly since it

Figure 3.6. Loaded van for a trip to dark sky country. (Photo courtesy Andrew Shulman.)

includes nighttime conditions. In the immediate 48 hours, an hour-by-hour forecast becomes available, and it has served me well, for the most part. This includes information on cloud cover, wind speed, and predicted temperatures. (The latter part of the forecast is more important than I can say!) The one aspect that is not always easy to deduce from this site is the sky transparency itself, but most of the time it is possible to have some idea from the overall weather descriptions.

Some time back, I stumbled across another Web site while looking for information on dark sky sites, specifically set up for astronomers in the United States and Canada. It features more specific viewing information for the immediate 48 hours than any other site I know. Recently featured in *Sky and Telescope* (April 2003), it seemed more than key to include it here, since it is something you should know about if you live in North America. It goes by the Internet name of www.cleardarksky.com and outlines, on an hourly basis, astronomically oriented conditions for a 48-hour window at hundreds of locations (many of them among the best dark sky sites). Regrettably, it is only in the United States and Canada. The difference between it and weather.com is that it expands the important base of viewing factors into actual astronomical viewing categories. These are cloud cover, transparency, seeing, and darkness of the sky. Thus, we now have a pretty comprehensive resource of information for our observing needs. If you live anywhere in North America, I recommend the site to be part of your own planning, and wish that I knew of a comparable site worldwide. Some other dark sky resources, including a couple in Europe, are listed in the Appendix.

Another part of the equation will involve the moon, which will be as significant as the weather itself in your planning. If the moon is in the sky to any degree other than a small crescent, you can forget that part of the night during which it is present, for any chance of attaining dark sky conditions. The Web site of the U.S. Naval Observatory provides an array of astronomical information, and a daily almanac may be accessed at http://aa.usno.navy.mil/data/docs/RS_OneDay.html. This is a worldwide site that can provide exact information of lunar phases, rising and setting times, and more, for any location. It is reported in your local time, a definite plus for many people who may be bothered by Universal Time conversions. Between these sources of information, or any others that you are able to find, some degree of preparation to reduce the chances of calamity becomes possible. This, together with your own common sense and judgment, should enable you to avoid disappointments most of the time. Your dark sky nights should be great occasions of joy, indeed the ultimate highlights in your entire viewing experience. This will be possible only as long as you can keep stressful distractions away!

Set Up in the Field

The ease of use of the little GVD-800 Walkman is a breeze! However, I would caution on carrying out some experimentation before using the setup on location, because the specific image brightness will be different on its own small monitor than on your computer at home. Even if your images look good on the monitor, you might not record them with appropriate brightness for later extraction to computer hard drive. It will be impossible to rectify the problem later, no matter how much you try to adjust for it during the process. I have found that the images may also need to be made to appear almost excessively bright at the time on the little 4 inch monitor of the GVD-800 in setting the video camera controls. You should bear in mind that they will probably not appear that way when transferred to computer. In my early use of this machine, I had some truly amazing planetary images captured on video, starkly and complexly detailed, only to find that they were beyond repair (too dark) for any type of satisfactory use, other than being viewed on the small monitor of the recorder itself. This was a lesson not lost on me, and now, having lowered the monitor's brightness to the minimum setting, I take great care to saturate any image to the maximum extent possible without burning out the brighter regions (another key issue). Other systems you may utilize will have their own considerations when it comes to image brightness. Be sure to "have a handle on it" before you spend hours recording the greatest sights of your observing life!

As for camera settings, generally speaking, virtually all image intensified deep space images require the slowest shutter speed possible, with sufficient gain and maximum contrast to secure full saturation. This is one of the most beneficial aspects of having a camera with a dedicated manual gain and contrast controls that are readily accessible. We always need to obtain images with the minimum gain necessary, in order to keep the images from becoming any more grainy or noisy than they need be, but not to the point of being insufficiently illuminated for transfer to computer. Therefore, work with your shutter speeds (in the case of image intensifiers, adjustable adapter iris) as well. Most of the time high gain settings are inevitable, but keeping them to the minimum needed should be an eternal quest, as long as the option is there. As far as I know, the Astrovid 2000 is one of the few cameras to have dedicated manual contrast (gamma) settings as well. This is a very important feature if we wish to make our images appear as close as possible to the live impression. Regardless of the imaging system you employ, I would also suggest, if your video recorder allows titles to be given to each subject during recording, utilizing this feature. The small extra time and trouble required may save confusion later and wasted imaging sessions, even though you may feel sure you will remember the identity of all of the objects you had taped at the time. (You will not!)

You should also consider the base of your telescope; we are not setting up in an area of developed land. It will usually consist of some form of dirt, dry and dusty, or worse, sometimes even wet and muddy, and usually sloping in one direction or another. I learned the hard way that 16 inch (40.6 cm) cement paving stones provide the best and most adjustable foundation for my particular scope. Nothing else proved stable enough, or possible to make sufficiently level. I usually make a four-pad support, digging them in and approximately leveling the placement of the paving stones by laying a straight edge across them and measuring with a simple spirit level. Because my home viewing deck is slightly sloped for water runoff, I also try to set the stones similarly, by noting the level's readings at home. I have found in practice, though, that exact leveling of the entire flat surface of each stone is not necessary, as long as the points of contact with the feet of the telescope's mounting are in about the right alignment. Your own telescope, no matter what size, would benefit from a similar plan of base setup. Even if you use a computerized altazimuth scope with full tracking capabilities, you still need a stable and fairly level base on which to set up, although the issue of polar alignment is often moot.

Before I proceed, I take care to set these paving stones as closely in line with the celestial pole as possible. This will save much frustration and time after dark. To accomplish this, I refer to my regular site at my home location, which I have meticulously marked on the concrete decking behind my house. For general viewing purposes on location, exact polar alignment is not necessary, although something pretty close is desirable. Precise alignment is not even particularly important for video imaging, although reasonable accuracy is again something that greatly simplifies the tracking process. To duplicate the alignment setup I have at home, you will need a compass with an adjustable true north position. Lining up a straight edge with two of the markers on my deck (a line joining the markers would be at right angles to the pole), a small table is set up, its edge aligned with the straight edge. (I need to do it this way, with a compass used on the tabletop, because my concrete deck is full of steel, which throws the compass off.) Now the table is polar aligned, at least in right ascension! I then set the compass on the table, and align one of its two straight sides (my compass has straight sides) with the polar-aligned edge of the table. True north may then be aligned with the true north of the compass, but not the table edge. Carefully adjusting the position of the magnetic north pointer, I can then replicate later the position of magnetic north relative to the table. It is thus simple to find true north at my dark sky site, almost exactly, before sunset, and possible to position and then level the cement paving stones quite accurately. The result is a usable alignment, achieved before darkness! You should allow a full hour and a half at your chosen remote location for this process, and more time if you want to further refine the polar alignment after nightfall. The issue of accurate polar alignment is likely to be more important with such video applications as the STV and StellaCam, since a greater degree of stability of the image is required for image building. Depending on the length of the exposure, or number of video frames being stacked at any given time, this need will only become more significant. (See Figures 3.7a and b.)

Before darkness, I think it also makes sense to try to have the telescope's collimation completed, just to save trouble and time. Some people prefer to wait for darkness in order to use the star testing method for final collimation, but I have always found that a combined Cheshire eyepiece and sight tube, if carefully applied, will produce exceptional results that seem hard to improve upon. Any use of the star test after dark has only confirmed my findings. (For example, Orion Telescopes of Santa Cruz, California, makes an excellent and very inexpensive combined Cheshire and sight tube that I personally use and recommend. See the Appendix.) My own telescope usually will survive being taken apart, driven to a remote site, and reassembled with virtually perfect collimation, having survived all of this without any adjustment whatever – quite a testament to its design and engineering. If anything needs to be done, it is of the most minimal nature only – something like 1/20 of a turn to one collimation bolt. (See Figures 3.8a and b.)

Having completed all of your preparations, you may still have a little time to sit down and relax, have a hot drink, and wait for nightfall. It's a grand time to spend time with friends. Try to make your circumstances as comfortable as possible.

a

b

Figure 3.7a,b. Setting up in the desert. (Photos courtesy Andrew Shulman.)

a

Figure 3.8a,b.
Collimation. (Photos courtesy Andrew Shulman.)

b

Bring lots of warm clothing, especially during the colder months, including protection for the feet, the first place of trouble. This is more important than it sounds; you are in conditions of unrelenting exposure to the night air, even if the temperature reading does not sound all that severe. You can't just go indoors to warm up by the fire. At your dark sky site, building a fire at the site will only interfere with the air and the image intensifier's need for total darkness; it may even be against park regulations. In an emergency, I know I can still retreat to the car and run the engine and heater for a while if I get really cold. Some good food, easily heated and served, does wonders too. (See Figure 3.9.)

Bring tools, flashlights (other than a red light, as long as you are not near other astronomers; your dark adaptation will be compromised anyway by image enhancement devices), magnifying lens (I could have used one when grit lodged in the connector plug!), and anything else that could help. It will not be possible to simply step into the house to lay your hands on anything you have forgotten, or when you need to adjust something, or simply switch on the lights. So, here is a suggested list of some key things to bring with you to your dark sky site (they are more important than you may think!):

Large flashlights.
Large standing red light.
Large camp lantern(s).

Figure 3.9. Dusk at desert site. (Photo courtesy Andrew Shulman.)

Lots of warm clothes, gloves, hats, thick socks, boots, and blankets.

Propane stove burner.

Radiating heater (such as a tabletop version of a standing propane heater; great for warmup periods, and instantly dissipating the heat they produce). Use away from your telescope!

Spare propane tanks.

At least two ways of creating a flame!

Lots of extra batteries, bulbs, and fuses for anything that may need them.

Small tables and chairs, preferably of the folding variety.

Plentiful and easy food to prepare (i.e., prepackaged, cooked, or canned).

Can opener!

Coffee, tea, or equivalent.

Brandy, if you care; a sip or two is a great morale booster on cold nights.

Energy bars.

Comfort foods.

Plenty of water for every purpose.

Plates, cups, knives and forks, utensils.

Cleaners for lenses and glasses.

Wide, soft camel-hair brush for cleaning optical/electrical components.

Magnifying glass (essential for delicate brush cleaning of intensifier components).

Pens, paper.

Spirit level.

Compass.

Shovel, trowel.

A watch.

General tools (essential for on-the-spot repairs and adjustments).

Medications you are currently taking.

I find I can completely fill a large van with all that I take; there is barely room for two people. All these "comfort items" really seem to count when you are out in the wild! However, there is one distinct advantage to freezing temperatures. Image intensifiers seem to produce ever-decreasing amounts of electronic noise as the temperature drops. Not only does this enhance our direct viewing, but it also can give us better video images with less frame averaging. It's just a small plus, but definitely worth the mention! Plan your viewing or video night very carefully; this may need to be different from the protocol you use at home. Have a schedule or catalog of some kind at your side, so that you can have instant reference to all you intend to see.

I consider it vital that any session in the wilderness be as stress free as possible. If you are imaging, you should allow yourself (yes!) to enjoy actually sitting down at a table by the telescope, basking in what you are seeing and recording on the monitor. Take time with each object, not only in viewing it in a leisurely fashion, but also in the care you take to set up everything for viewing it. Do not allow these sessions to become marathon scramblers! Sometimes, less really is more! I also take the time to move the table with equipment and the chairs, between viewing objects more than a few degrees apart, whenever it makes life even a little simpler. There is nothing worse than struggling to stretch electrical cords, and more likely, your patience! The whole operation will likely become very trying and unproductive if any manifestation of chaos or frantic pace is part of the proceedings. Believe me, I've been there.

Cleaning Optical Components

A couple of final considerations are regarding the age-old issue of the necessity for clean optical components. Do not wait to get to your remote site, only to find that key components of your equipment are dirty. The bottom line is that for true live viewing, only one piece of equipment needs to be scrupulously free from dirt of any kind: our image intensifier, and specifically the phosphor screen. If the slightest speck lands on this part, you can expect it to show as a sizable black dot in the image. In normal circumstances the phosphor screen is internal in the unit because the eyepiece component seals it from the air, and it is not subject to particles in the air. However, in setting up for video, the eyepiece will be need to be removed, leaving the phosphor screen open to the air for a brief time. Little particles of grit, even metal shavings from the screw thread itself, can fall onto the screen during this time, and are maybe even attracted to it. When the screen does require cleaning, the greatest care needs to be exercised when cleaning it; usually a camel-hair brush, a magnifying glass, and a bright light is about all that is advisable, with maybe some gentle help from lens-cleaning tissues.

In any video application, dirt becomes a far bigger problem in other elements of the optical/video train. Submicroscopic specks on the CCD video camera's chip face are even more greatly magnified than on an intensifier's, since the area is so small. Use of powerful Barlows (i.e., 3X and above) makes the most microscopic particles hugely visible, appearing almost as boulders on the image! Camel-hair brushes are not usually sufficient in such extreme cases, as they tend to deposit minute particles of their own. I find that many repeated attempts with proper lens cleaners and papers, followed by careful inspection with a magnifying lens under a bright light, is about the only way to succeed. The exact method I employ is somewhat tedious but seems to be the only way that is totally satisfactory. Briefly, what I do is as follows:

Multiple attempts are made, each attempt consisting of spraying a small mist of cleaner, followed by a single, one-directional wipe of the tissue; it seems to be the best method for ultimately removing particles, and the slight moisture helps to make particles adhere to the paper. Many tries eventually succeed. Other motions in different directions will only deposit anything we have previously removed. Use pressure sparingly! This procedure can actually occupy hours when one is striving for perfection! It is extremely tedious, but will be worth the effort.

With regard to cleaning the primary and secondary mirrors of Newtonians and the like, for the most part they have a considerable tolerance for grime. While not meaning to encourage dirty optics(!), these mirrors will perform quite well even when they appear a lot less than pristine. If your primary seems to have gone beyond this point, or you are simply interested in obtaining maximum reflectivity and minimum light scatter, I would like to share a trick I have utilized over the years. Although the procedures prescribed in the usual manuals on telescopes work well when used with experienced hands, they often produce great frustration in less-adept hands when it comes to completing the job for a spot-free, truly clean surface. So you might try the following method, for a fast and easy yet very effective and safe mirror cleaning procedure:

Place the mirror in a large sink, tub, or shower stall. Do not submerge the mirror in water so that particles of grit can come to rest on the mirror face, causing probable scratching of the delicate surface as it is cleaned. Instead, using a flexible hose and spray (either the kitchen-sink variety or hand-held shower unit), spray directly across the mirror face from several directions for a minute or so. This will knock off the most abrasive particles, while not adding any new ones. Now mix some liquid detergent with warm water in a large jug, and swill it over the face of the mirror. Let it sit for a couple of minutes. Take a large cotton swab and, using no direct pressure other than the weight of the swab itself, work it over the face of the mirror in large circular motions, working from the center outwards. This method is completely safe because of the method we have taken to eliminate large, abrasive particles, and the lack of applied weight to the mirror surface. You might change the swab a couple of times during the procedure for a larger mirror. Spray clean and repeat the process. Finally, having sprayed the surface clean of all soapy water, stand the mirror at a steep angle. (Do not let it slip!)

Now comes the most important part. Since avoiding water droplets and spotting has always been the main problem we encounter, it might seem there is no ready way to keep them away, but there is! You will not need distilled water, blotters, alcohol solutions, special skills, or anything else to accomplish this. Simply take a clean, wide-mouthed jug (such as a glass measuring jug), and fill it with fresh water. Rest the side of it, away from the mouth itself, against the beveled edge of the mirror (yes, touching it) at the farthest part of the mirror from you, and begin to pour the water sideways in a wide stream over the mirror face. Slide the jug around the top half circumference of the mirror as you do so, and as long as you maintain a wide, steady stream, and keep contact with the mirror edge, you will find that not a single droplet will form! Amazing! The whole cleaning procedure takes minutes and produces outstanding results.

After We Have Survived a Night in the Wild

Once home, the process of obtaining good still images (frames) is quite straightforward. Remember though, if you have used an image intensifier and your still images are to make any sense visually, the deep space video still needs to be processed through a recursive frame averager, and then fed to the computer, in much the same way as when recording live images. We will save space in our computer's memory if we store only those video clips, or portions thereof, that we will ultimately use. Playing these complete video clips from the analog output of the VCR, the signals are fed through the recursive frame averager. Usually, I average 16 frames at a time in deep space imaging. Although the resulting image is a little "softer" than the original live view, it is easily the most pleasing when it comes to extracting still frames. From here, the procedure I use is to reconvert the signal to digital code, via the Sony Media Converter. Then it is fed into the hard drive via "Firewire" of my iMac computer through iMovie software, a very simple and straightforward system. It is quite easy to view and download the video clips on the computer monitor. The net result ends up being the same as doing the entire process at the time of recording, except it is now a two-stage operation instead of one. There seems to be no downside.

A video recorder such as the GVD-800 also allows each frame (or averaged frames) to be scrutinized and selected. If you use moving video software, in whatever form that may be, it should also allow you to "freeze" or view individual frames, as with the video recorder. When you select individual frames, even after averaging or compounding, you will become aware that not all frames are created equal; what we see in the moving image is a composite of many slightly differing manifestations of the same object, to create the whole. The frame we will ultimately select needs to include as many of these fragments of visual information as possible. You may be surprised how difficult that sometimes is, but the chase is always fascinating and worthwhile. In certain still frames, the finest details suddenly jump out and stare at you.

The only downside of recording video on site, without frame averaging, is the less than satisfactory, slightly snowy images at the time of their making. One also has to make all judgments based on much less than optimum visuals, but after becoming used to the procedure it is not hard to produce outstanding results. The main issue is to obtain the best focus possible; take the maximum trouble to do this, as it will repay you in spades later. Some degree of poor contrast or brightness can be adjusted at another time, but not focus. However, at home, what comes across on the monitor will likely not equal the effective contrast or crispness you saw with your own eyes at the time. This is true with any type of live enhanced viewing. The human eye is a complex light detector, registering images with unique sensitivity.

So now I know you are wondering, "What can I expect to see with these new methods and approaches?" Let's go on to Chapter 4.

CHAPTER FOUR

What Can We Expect to See?

Seeing deep space objects live with conventional viewing is something of an acquired skill; while the challenges are significantly less with an enhanced viewing device, such techniques remain needed even then. So while the effect of enhanced viewing can even be quite representative of the grand images we are used to seeing, we will nevertheless have to learn how to extract such imagery from the still often vaguer and fainter live image that actually meets our eyes. Once the eye and mind have made their own adjustments, the differences in brightness and contrast become less noticeable. Enhanced viewing also has the distinct advantage of being able to involve a group of people, who may well be inexperienced observers, to show them deep space wonders with much greater ease than might otherwise be the case. We have all experienced the frustration of novice observers as they struggle to see what is plain enough to our eyes.

If you are using an image intensifier with monitor viewing for a larger group to observe simultaneously, the little viewing screen of the Sony Video Walkman recorder, or better, some other larger high-resolution monitor will show unprocessed images quite well. However, for this purpose it may pay you to provide recursive frame averaging as well at your remote site for the best results. This means more electrical and wiring problems (sorry). Luckily, it's not an issue with the other integrating video devices I've described.

Time exposures in any form (photographic or CCD) remain an entirely different form of viewing. It is, of course, possible to image faint objects with great results utilizing these techniques, but it is not possible to witness an object at the time. I would not want to suggest that the results obtainable using the simple direct video imaging I employ are equal to fine CCD imaging, although occasionally it does approximate the effect surprisingly near the mark. (Though the same holds true for any form of video, the STV would seem to come closest of all.)

However, we must not lose sight of the fact that "viewing" by CCD imaging is, by default, nothing like the old-fashioned visual approach, which packs its own unique punch. If we must have something for the record, the form of imaging most appealing to me is real-time video, but ultimately it serves only as a means to an end – that of preserving a reminder of my own live viewing experiences. I cannot deny that having something to bring home after an expedition to the dark skies is something of a trophy; it is wonderful to have images to remind you of the wonders you saw with your own eyes, even if they do fall short of the live experience.

So with all of these unequal factors, what can you expect to see at a dark sky site? With only moderate apertures, it is possible to view detail and actual spiral structure in many galaxies with relative ease, and often without using averted vision either. Compared to conventional viewing, dark lanes and emission regions within favorably placed galaxies also can often become stunning. Many of our own galaxy's emission and planetary nebulae become bright and full of intricate detail, totally unseen in conventional viewing with the amateur's telescope; star clusters, especially the globular variety, explode with bright, resolved stellar points and dark lanes, and can appear easily resolved down to the core. These are the most dazzling of subjects in the enhanced view.

The seasoned observer is accustomed to extracting detail from faint subjects, using advanced viewing skills with conventional equipment. However, for those subjects to which it is well matched, an image enhancement device makes the job much easier. Along with potential detail and possible spectacular revelations, enhanced viewing provides an opportunity for seeing many otherwise faint objects brightly illuminated, unlike what we are accustomed to seeing in deep space viewing through a conventional eyepiece. Just to whet your appetite and give you some idea of just what is possible, Figures 4.1a–e present a few views of deep space subjects, single frame video images taken with my 18 inch (46 cm) telescope and my image intensifier and looking remarkably akin to their live appearance.

None of these examples are particularly successful candidates for image enhancement in the suburbs, so it should be clearer by now the potential I am

Figure 4.1a. M51 The Whirlpool Galaxy.

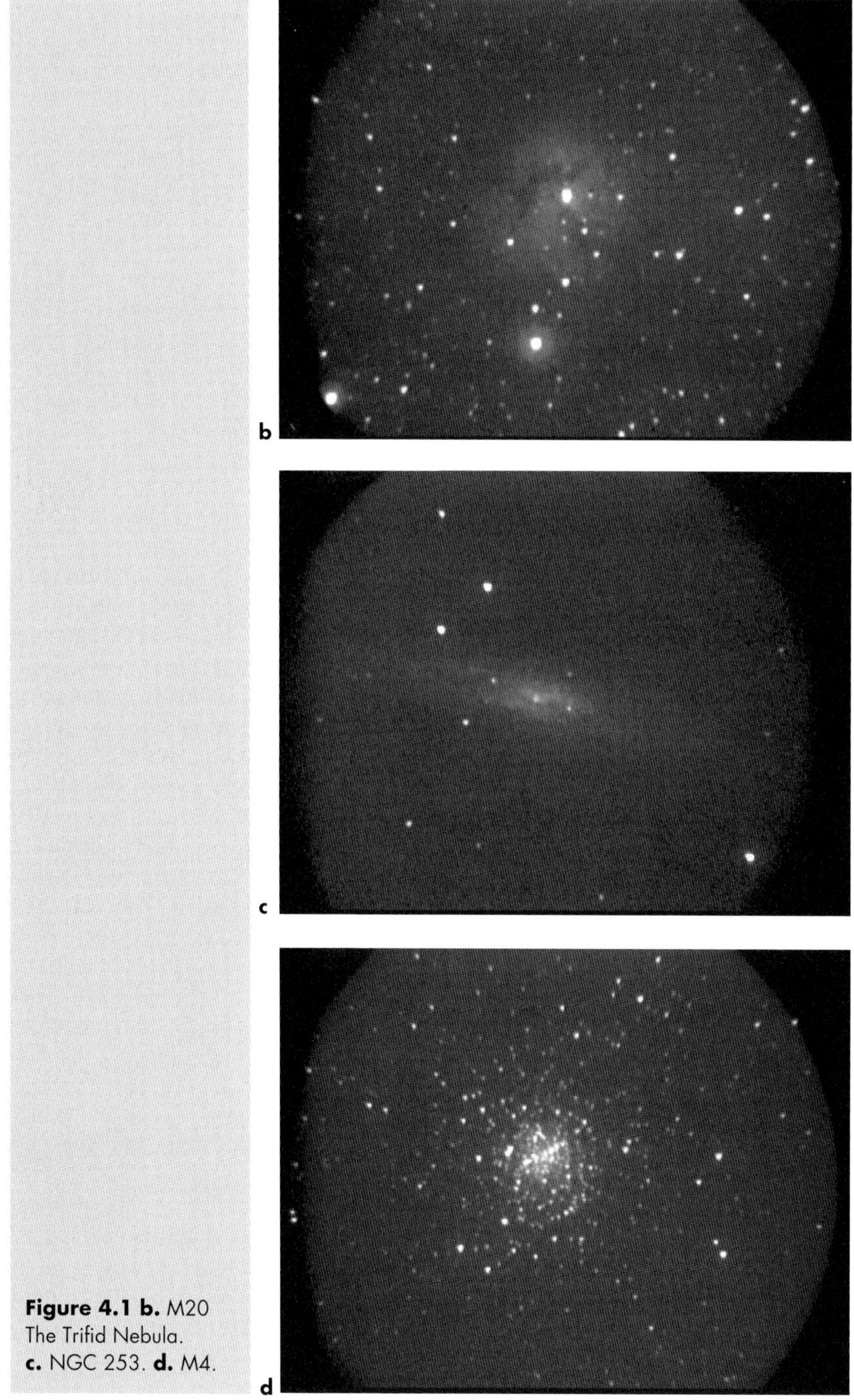

Figure 4.1 b. M20 The Trifid Nebula. **c.** NGC 253. **d.** M4.

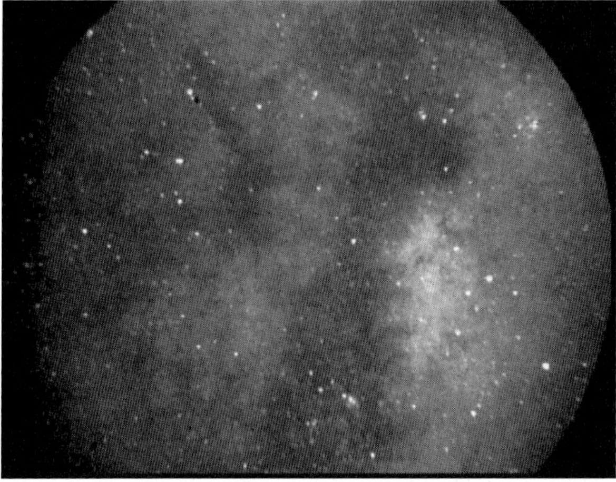

Figure 4.1e. The Pipe Nebula and Milky Way Hub.

trying illustrate in hauling our gear to dark sky country. While it is also true that I use a sizable telescope by many standards of the amateur's universe, the best astronomical image intensifier available, and have access to outstanding dark skies, I can still promise you quite a ride with something less than the biggest and best available. There are reasons for this. As you know, the lower magnifications (and hence smaller, more concentrated images) of lesser apertures often will make up for some of their reduced light grasp. While this is not the case in their resolution, it is the case in their basic ability to show objects in space with greater ease. Second, telescopic size is not always the answer to seeing better. As we go ever larger in aperture, we find that some of the more sizable deep space objects are less suited to live viewing. This is because the minimum magnification we can use with these "light buckets" is too great for the eye to register some of the more extended and diffuse subjects effectively. Aperture alone does not necessarily compensate sufficiently for the eye's inability to compound photons in live viewing. Certainly processed images taken through advanced integrating video cameras can be very bright, smooth, and impressive, such as the examples of images in Figure 4.2a-d taken through relatively modest apertures (indicated) with a StellaCam EX frame integrating video camera (by James L. Ferreira, courtesy Adirondack Video Astronomy).

Deep Space Reference Materials

For reference sources, there are numerous publications available, all dealing with the amateur observer's specific needs. I have always been an avid devotee of Burnham's *Celestial Handbook* (Dover Publications). Those of you who have read *Visual Astronomy in the Suburbs* already are well aware of this! Burnham's three epic volumes (still reproduced exactly as the author prepared them on a simple typewriter, prewired processor!), lovingly and meticulously written and always eloquently informative, leaves one to marvel just how one man was able create

Figure 4.2 a. M56 (25 cm). **b.** NGC 6826 (25 cm). **c.** M82 (15 cm).

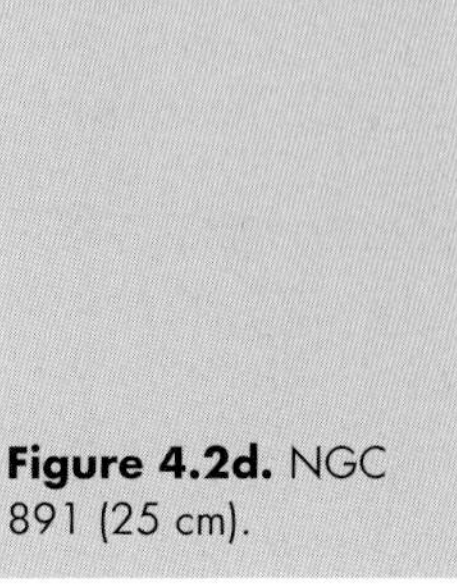

Figure 4.2d. NGC 891 (25 cm).

such an amazing resource in his spare time. It remains quite inconceivable to me. Burnham's will likely always occupy hallowed ground, and will likely remain a unique and seemingly unsurpassable high point in reference and literature for the amateur astronomer. However, even within these colossal volumes, there simply is not room to provide detailed observing guidance to all of the objects it lists, or even detailed descriptions of most. It was intended to be an information warehouse for the amateur observer concerning many facets of scientific discovery and observation of deep space. Thus the narrative viewing guide to its many cataloged sights is only part of the whole.

In days when the size of amateur instruments was typically much less than those that are commonplace today, Robert Burnham's books provided a mine of information for most observers. This is especially the case when we realize that any amateur possessing a telescope larger than 12 inch (30 cm) aperture was an astronomical rarity indeed. While Burnham listed basic information on all objects viewable in telescopes up to 12 inches (as seen under dark sky conditions), he provided detailed notes on only the more significant ones. In truth, most of his readers' own telescopes were probably not much more than half this size, and therefore this approach made complete sense. As a resource for this present book, Burnham's could probably still have sufficed, since I am still not commenting on deep space objects that may be less than spectacularly revealed. (Objects that a 12 inch or larger telescope could merely detect or partially reveal are not the stuff of my own writing anyway.) Those objects that are less than completely examined and discussed in Burnham's would not have dealt the project any fatal blows. However, looking for further published resources resulted in an excellent find, more complete as a reference for individual deep space objects than any other I have uncovered:

The Night Observer's Guide (Willmann-Bell, Inc.), a massive two-volume set, by George Robert Kepple and Glen W. Scanner, turned out to be that new resource. Often touted as a modern-day Burnham's, I should point out that its purpose and execution is entirely different, and it is certainly not comparable as a literary or similarly substantive work. Just reading the clumsy prose of the preface to Volume 2 demonstrates this amply! However, it is important to understand the difference between the two approaches. *The Night Observer's Guide*, while leaving alone the often profound and personal insights that Robert Burnham labored to share with us, does set out to address some special practical needs that today's amateurs have. In being centered around actual observations in the field of countless deep space objects, and not so much the scientific and historical backgrounds surrounding each object, it is a valuable reference. However, its real strength is that it allows us more planning in advance of the deep space objects we may wish to spend our observing sessions locating and viewing. The vast array of photographs and drawings found in this recent guide are also by amateur astronomers, all from actual observations and recorded images. Good as they are, generally they are not of the order of most observatory or modern CCD images, presumably part of the authors' thinking in not promoting unrealistic visual expectations (a wise idea).

Actually, the *Guide* does not generally explore deep space to significantly fainter magnitudes than can be found listed in Burnham's. This is among the reasons why I certainly do not accept the *Guide*'s premise that Burnham's is limited to telescopes of up to 6 or 8 inches only! Despite some surprising errors in the *Guide*'s opening essays, coupled with many ill-written sentences and "typos" throughout its two volumes, *The Night Observer's Guide* has become an important part of my preparations for observing sessions. In these volumes you will also have to forgive some unexpected omissions, as well as needed commentary about various well-known observing challenges, such as details and plotting references to the fainter stars in the Huygenian Region of the Great Nebula in Orion M42; we could have used them. Instead, for such detailed background or information, you will find yourself returning to Burnham again and again. And guess what? Usually you will find what you are looking for!

More startling in the *Guide*, however, is the absence of numerous Southern Hemisphere constellations, and the significant deep space objects contained within them. Some of these are among the greatest sights in the entire sky. These gaping holes go without even a cursory remark or explanation, which we can only deduce is because all of the contributors to the books are located in the Northern Hemisphere! While partly understandable from the perspective of these northern observers, it must be very irritating and frustrating to those in the south, especially in finding a set of books that appear, at first glance, to be complete guides. Inexplicable as some of these issues may be, this recent and epic resource nevertheless should probably figure large in any serious observer's book collection. It is still the best example I have seen of a practical, in-the-field, amateur observer's reference.

This brings us to an appraisal of the purpose of this particular book that you are presently reading. Where does all of this place it? I hope it will not be judged by the same standards I may appear to be imposing on others! I did not intend this volume to serve as a complete guide or listing of deep space objects to view. Instead, it is my hope that it will reveal the huge potential that image enhancement technology can open up to us. Maybe it will whet the appetite of those passionate about live viewing with just what can be done by taking advantage of the right equipment and circumstances.

Now Let's Get to It!

The succeeding chapters are divided into categories of deep space objects, described and illustrated accordingly. Since these chapters form the mainstays of the stated purpose this book, they are few, numerically speaking, but often feature a proportionately large number of pages in each. Not surprisingly, the chapter on galaxies ranks as one of the larger among them, galaxies being, after all, the grandest and most defining objects that make up the visible universe. However, although I have included a wide representation of each category of deep space objects in these pages, I do not mean to imply that my selections are in any way a complete catalog or listing. They are meant only to show what can be done using the new equipment available, perhaps even to inspire others, and above all to give a reasonable perspective for expectations of the live viewing experience you may be about to have. You will notice that my selections include some familiar sights, often selected from among those that exhibited significant gains in visibility from suburban sites utilizing the same approach. Comparisons between the two are often startling; now that we will be observing from a dark sky site, we can probe ever deeper into these and many others in the night sky. Hopefully, you will find encouraging material for the spectacular live viewing that you, like me, may crave.

Unlike in *Visual Astronomy in the Suburbs*, I decided against including drawings of the space subjects this time around, as the advantages of moving to a dark sky site make realizations by hand less valuable and more subtle. The video frames, and certainly those I myself recorded via image intensifier, are intended to indicate approximately what may be seen live. This is something I was seldom able to achieve from the suburbs without drawing the astronomical sight in question as well, in order to give a more complete representation of the whole. Although one can more easily convey the crispness, resolution, and brilliance of the live views by drawing, in the video images that now become possible we can usually appreciate most of the distinctions that separate dark sky sites from the rest. Generally the qualities that I describe show in the still video frames. In many instances, the scale of the images is relatively modest, providing the greatest image brightness, and corresponding more closely to live enhanced views experienced; many of the video frames were made with the lowest power. It was also difficult sometimes, when larger scales were selected, to suggest the visual effect and impact many deep space objects have on the eye in live viewing, even when these objects were originally seen well at these higher powers. Somehow, many of these looked too diffuse on the page, lacking the presence of the real-time view.

I also did not necessarily exclude any image taken under less than perfect conditions. The advantages of a dark sky site are so significantly better than our suburban surroundings that it is not always necessary to have perfection to achieve decent visuals. I hope the images reproduced here (starting in this chapter) will therefore provide more reason to be optimistic for those frustrated by less than fully cooperative skies; we have all seen too many illustrations taken in superlative conditions that we seldom experience for ourselves. The real world is seldom like that for most of us. Sometimes we are simply grateful if it just does not blow, cloud over, or rain! Again, I am not so much concerned with being able to simply discern a subject, or making out slight and grudging detail; just the detection of these objects is not enough by itself for my purposes here. Therefore the so-called "faint fuzzes," an important part of real-time observing to all amateur observers, nevertheless do not represent my largest driving motivation for this book. However, even "fuzzies" can be significantly enhanced by the methods espoused in this writing. Indeed, with an image intensifier or advanced video system, the increased potential for detecting deep space objects far fainter than those featured here is a given; many of these may be considerably fainter than the amateur is accustomed to viewing. I should say though that, unless these objects are clustered together in great bunches (such as the Fornax or Virgo Clusters), the potential for spectacular live viewing of such objects only goes so far.

Addicted as I am to the specter of seeing a deep space object stunningly revealed, and in real time, I always hope to witness deep space in a manner more reminiscent of the images we have been accustomed to seeing from the great observatories. Sometimes this is quite attainable on many of the most suitable subjects, even without any of the new special aids, and especially in our new surroundings. It is equally true that some deep space objects are less likely to lend themselves so well to image enhancement, and more likely to reveal their presence, or details, with conventional viewing instead. Many subjects that are less well suited to enhancement are only revealed as brighter, revealing no increase in visible detail when viewed in this manner. Again, this is because these subjects' spectrums are less suited to enhancing devices, or the delicacy of the subject tends to be obliterated in the electronic image. Others become less visible or even invisible. We will leave those less-suited sights to other existing volumes.

Luckily, in many instances, only enhanced viewing makes the objects appear truly spectacular, at least in the sense of approaching the impact of a time exposure. Still others present themselves to us in equally valid ways with either form of viewing. The great galaxy M33 in Triangulum is a good example. With a conventional eyepiece, its familiar shape and surprising detail is quite apparent, although one has to learn to increase eye and brain sensitivity to faint subjects. Even in the suburbs I have been able to glimpse its spiral arms on transparent nights; though very faint, its outline and patchiness are sometimes vaguely revealed. However, it also much more successful when viewed with enhancing devices from dark skies; this cannot be claimed in the suburbs, as the galaxy disappears completely from view. Dark sky sites give us new possibilities with our new toys, unknown near the city. The results of enhanced viewing have astounded me so often that I anticipate that it will bring similar results for you, although you will be aware of a reduction of some of the subtleties we saw conventionally. The key is to know what type of subject is most likely to benefit, and you will soon acquire a feel for those that are more likely to respond favorably, although you never know for sure. On the many that do, enhanced viewing comes into its own, and many aspects of an object are stunningly revealed, often for the first time. However, there is still nothing quite like the purity and exquisite refinement of the natural images of enlightened conventional viewing, with any type of deep space object.

Because I am not concerned with pursuing actual ongoing research itself, I will admit that this is not my driving motivation. It sometimes feels as if surely I must be some sort of astronomical heathen for merely appreciating and enjoying a visual feast only, more than the subtle detection and detailed study or research of many faint, esoteric, and sometimes near-invisible objects! However, I am quite sure I am not alone; maybe some are ashamed to admit to such unfashionable or apparently nonintellectual sentiments. There should be no shame in the informed pursuit and enjoyment of all of those spectacles we can find in the universe, regardless of what some might say! Meanwhile, there should also be nothing but praise for those who wish to pursue a higher, scientific ideal; these are the folks who will be most likely to move into a professional role, or those who cross between both worlds. Without them, we wouldn't know what to look at, or what their findings represent in the cosmos.

This book will have more than served its purpose if it is able to convince some conventional visual observers, otherwise resistant to some of the new technologies, to be open to the huge possibilities of some other new approaches to their real-time viewing. After all, should not we all be primarily concerned with expanding our viewing potential? I will also be more than satisfied if some newcomers to astronomy are inspired to try this alternative approach in tandem with traditional viewing, even from the start. I can see no reason why they should not do so, except they should know not to neglect the more conventional viewing methods. Visual astronomy can further add to our experience and understanding, but only when built on a solid foundation, including all that conventional viewing has to offer. Enhanced viewing is not meant to supplant and replace this time-honored tradition, with the vast range of subtleties and refinement only it can provide, but to add a new dimension.

Providing a wide cross section of deep space objects that may be seen anew and to spectacular effect, it is my hope that the value of "the new way" can provide a fresh inspiration; it is also my hope that my book may provide some realistic expectations. It is now time to relocate to a place of favorable viewing, armed with our powerful new tools of real-time viewing, in order to begin again our discovery of the night sky. Hopefully, this new journey into deep space will be something of a revelation, and perhaps even bring about another level of enjoyment. And maybe finally a time will come when amateur astronomers also finally embrace the amazing green glowing tube.

CHAPTER FIVE

Our Neighborhood in Space: The Milky Way

Our local neighborhood comprises almost all that we see with our unaided eyes when we peer out into space at night. This includes all of the stars, which include that great glowing band known as the Milky Way, and the occasional blotch of cosmic haziness likely to be nebulae or star clusters. This is our own galactic backyard; almost everything belongs to our local system, the Milky Way Galaxy. The only exceptions could be counted on one hand. Through the telescope, our backyard opens up to reveal a dazzling array of wonders, and so we will begin here with our new viewing devices.

Nebulae

Unimaginably vast clouds of billowing gas expanding far into interstellar space, sometimes illuminated or lit by the radiation of stars – so we could describe galactic nebulae. The Milky Way Galaxy, like most spiral galaxies, is literally studded with these in a wide assortment of types, origin, and variety of appearance. Those within our own local galaxy represent great potential subjects for amateur astronomers, and may be more interesting to some observers than anything else in deep space. Certain nebulae are the birthplaces of stars; others represent the end of stars' lives, even their graveyards or ghostly remains. Many of them give us some of our best live viewing opportunities. However, dark skies are needed for most varieties to show themselves to best advantage, and image enhancement does not work equally well on all of them. However, on suitable subjects the results are often spectacular indeed, with the appearance of many of the nebulae being more brilliantly illuminated than anything you may be prepared for, even when they are viewed from the suburbs. Let's look at the various types of nebulae, and examine some prime examples, as well as their suitability for enhanced viewing.

Reflection and Emission Varieties

These frequently awesome creations are likely to be stellar nurseries, and we see them via the light emitted by their offspring, breathing their own new life deep within them. With reflection nebulae, what we actually see may be merely reflected light, emitted by stars within the gas clouds themselves. This is the lot of reflection nebulae, only seen as a favor granted us indirectly. This reflected light normally consists of blue wavelengths, and as you already know, this is the very part of the spectrum that is the Generation III image intensifier's weakest response. Under such image intensification in particular, reflection nebulae are often quite disappointing, but maybe here Generation II devices have their greatest potential value. Video cameras of the type I have outlined may do better still with some subjects, but they too have a different response to the eye. However, I have sometimes seen some quite impressive results obtained with these cameras. Specific results will depend on the precise frequency response of the camera utilized (they may also emphasize the red and infrared end of the spectrum), as well as the specifics of the nebulae themselves.

Most of the time, it would seem that reflection nebulae are best seen with conventional viewing, using narrowband light filters and low powers. However, even when a given reflection nebula proves to be a flop with our enhancement devices (which is much of the time), it can sometimes reveal myriad young stars buried within it instead. This is justification for our efforts in itself. Better still is when it exposes regions of the cloud that are actually of the emission variety, and it is here that these reflection nebulae may become particularly interesting for us. We may greatly benefit by seeing enhanced regions of the nebula, due to that portion of the light spectrum not otherwise prominent, or even invisible to the unaided eye. In pure emission nebulae, the energy radiated from the imbedded stars is such that it excites the nebulae into fluorescence. Often the light of these nebulae, or emission components of what are otherwise reflection nebulae, fall within the red and infrared portion of the spectrum. They are likely to be prime candidates for enhancement.

From a visual standpoint, we should therefore differentiate carefully between these two varieties of nebula, since it is with emission nebulae, or emission components of a larger reflection nebula, that we will have our greatest potential with our new tools. But perhaps the most striking attribute that often becomes possible in enhanced viewing is the sheer brightness of many emission nebulae, which otherwise might appear ordinary. It is also true that certain nebulae of this variety are so radiant in illumination that they reveal themselves in dazzling ways with both enhanced and conventional viewing. However, enhanced viewing imparts an entirely new dimension to the live experience. I will not say that it necessarily surpasses the conventional view, although it can indeed do so, but what it does show is entirely different, and adds a new level of enjoyment and understanding.

Results in the Field

Perhaps the most stunning example of what is both a reflection and emission nebula is the Great Nebula in Orion M42, relatively nearby in our Milky Way's larger galactic terms. From a dark sky site, with any form of viewing, the wonders of this nebula are even more pronounced, its outstretched "arms" seeming to reach far beyond the visual boundaries, like a vast pterodactyl soaring through space. These arms seem to be trying to curl around the center of the nebula itself, like a huge bear hug. In the field of view the sight is a jaw dropper! Interestingly, and not at all surprisingly, the view is differently presented to us under image intensification, as extreme ionization from grand-scale star formation within the central Huygenian Region becomes newly and brilliantly emphasized.

The new Collins video adapter for the I_3 image intensifier finally makes it possible to show or capture on video the entire sweep of the normally visible "wingspan" of M42, even with a telescope of as great a focal length as mine (81"/206 cm). This particular nebula is a vast blaze of light, offering many possible video exposures. By this I mean that the iris of the adapter can be variously adjusted to show the stellar makeup of what lies within the nebula, or can be increased in increments to saturate the image with the subtlest parts of the nebula. All of the options are viable, depending on which aspect you wish to emphasize. You will need to decide for yourself the exact degree of brightness you should select; it may be necessary to swamp the central region in order to better show the peripheral extent of the nebula. I prefer to compromise somewhere in the middle in order to preserve some of the detail of the central region. There is so much light reaching us from M42 that in live enhanced viewing it will light up the immediate area around the telescope, and ruin any dark adaptation you may have attained! Here is surely one of the greatest marvels of the sky, dazzlingly revealed anew.

Actually, the Great Nebula in Orion consists of two primary separate parts, separately numbered and cataloged by Herschel. The smaller component, M43, a comma-shaped nebulosity with a bright central illuminating star, lies across from a dark bay and wide channel (sometimes known as the "fish mouth"). Not unlike the "lagoon" in the famous Lagoon Nebula M8 in some respects, this is a true dark nebula component of the whole. (See the subsection "Dark Nebulae.") It actually shows darker than the surrounding space around the great nebula itself, a feature easy to see in the first video image. We have to bear in mind that most of the entire region surrounding the great nebula is to a greater or lesser degree illuminated by nebulosity, even when its effect is slight. Because of this, the dark nebulous areas look extremely dark. (See Figure 5.1.)

However, the most significant portion visually of the entire spectacle is the Huygenian Region, the brightest part of the whole, and known as a giant birthplace of stars. Under intensification or suitable enhancement, the prominent emission gases of the Huygenian Region become so bright the view will easily stand high power. You can use a seemingly outrageous amount of magnification, should you wish, in order to see ever-more-refined detail, although you will be seeing ever-smaller segments of the whole. (See Figure 5.2.)

The closeup image of my own here shows even more clearly, in addition to all the benefits of dark skies and transparent air, the greatly enhanced clarity and

Figure 5.1. The Great Nebula in Orion M42, by James L. Ferreira using a StellaCam EX camera through only 80 mm aperture. (Courtesy Adirondack Video Astronomy.)

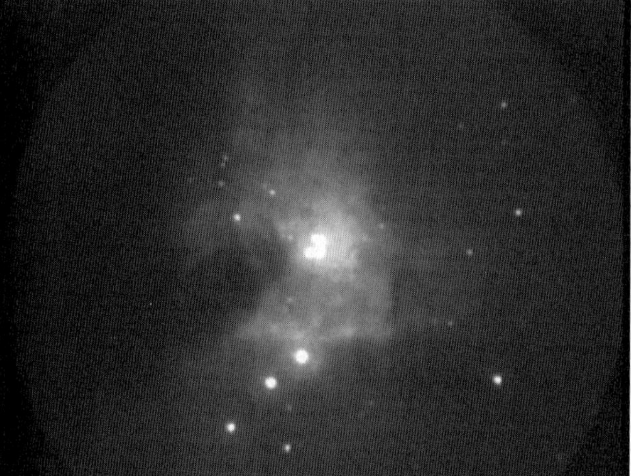

Figure 5.2. Huygenian Region in M42.

definition of the improved Collins video adapter. You might compare the result against that which I was able to achieve in my previous book, *Visual Astronomy in the Suburbs*. In the image here, intricate detail is revealed hardly less impressively than in some of the most imposing photographs we are likely to see. The Trapezium, the remarkable multiple star, is only part of an approximately 300-member stellar group, a very young star cluster in the middle of formation. Possibly the most celebrated star factory known, image intensification makes many faint stellar members of this group possible or much easier to see. You will see more of these in the live view or video than you are likely to discern in a still frame. It may be of interest to plot those stars you can make out in your own observing sessions from detailed references. (See, for example, Burnham's for detailed information and star charts of the entire region.) More power than I have utilized in the illustrations here is well within the grasp of my combined I_3 and telescope on this particular subject!

Figure 5.3. The Lagoon Nebula M8.

The Lagoon Nebula M8 in Sagittarius is another of the great showpieces of the skies, and is breathtakingly beautiful even from the city; nothing quite matches its mystical glow and sparkling embedded gems (Figure 5.3). Another emission as well as a reflection nebula, the full extent of the gaseous mass containing M8 is actually far greater than the portion of it observable. It actually extends to and merges with the Trifid Nebula M20 nearby, though it is well separated in telescopic terms; in fact, this entire region of Sagittarius comprises one huge nebula. However, we are fortunate that at least parts of it have been made visible by the energy of stars buried deep within it. The visible region forming what we see as M8 is another vast HII region of star formation in our galaxy. Buried deep within the eastern portion is the hot open star cluster NGC 6530, which sparkles like a jewel box. The most energized stellar components of this group provide most of the radiation that illuminates this side of the nebula. However, and perhaps surprisingly, this is the least striking region of luminescence among the whole, but it is the part that glistens with multitudes of sparkling points and is the most magical component of the whole.

The brightest portion of the nebula lies on the other side of the great dark lane slashing M8's midsection (the reason for the name "Lagoon Nebula" in the first place). Some might argue that this dark lane resembles less a lagoon than it does a rapid stream or torrent. Dark and bright streaks within the channel and parallel to its length also may be seen within this "lagoon" region, which reminds us in some ways of the similarly flowing dark sections of M42 adjacent to the Huygenian Region. Other lanes and blotches are to be seen swirling all around the vicinity of the entire nebula, some being quite broad, but less defined or apparent than the lagoon. Once we wade across the stream to gaze at the eastern banks and beyond, we see that it is substantially more brightly illuminated (by embedded stellar fluorescence), but more specifically, the magnitude 6 star, Sagittarius 9.

The Lagoon Nebula's emission and reflection makeup will become clear when observed under intensification or other red- and infrared-sensitive devices. It is then that the contrast between what you will see with conventional viewing and

Figure 5.4. The "Hourglass Region" in M8.

image enhancement is most striking, as parts of the nebula seem to simply disappear when viewed in the electronic images, while others are brightened. However, in our dark sky surroundings, newly prominent regions, of which we may have been previously unaware, are suddenly apparent. Most notably, the "Hourglass Region" to the east, challenging to see as a distinct outline with conventional viewing, becomes readily visible when seen with my I_3 (Figure 5.4). Its hallmark outline makes the reason for its name clear at once, and with sufficient power, it shows now as the brightest and best-defined region of the whole nebula. This feature is not commonly described by observers who have viewed M8 with conventional methods and moderate or even larger telescope apertures only. It is indeed a great bonus that we are finally able to see Herschel's aptly described feature so starkly revealed, and in real time at that; the hourglass itself even shows from locations far less suitable. Notice the two dark lanes extending northeast from the hourglass, appearing almost like a pair of rabbit ears. This part of the nebula will withstand some quite considerable magnification.

Although the Lagoon Nebula is connected by nonilluminated gas to the Trifid Nebula M20, this is considered a separate entity and is another one of the most famous sights in the sky; it is also a place of ongoing creation (Figure 5.5). Often providing the inspiration for countless artists' generic illustrations of faraway places in deep space, the Trifid certainly lives up to its mystical-sounding name. This is the case in almost any circumstances with sufficient aperture, including even conditions of substantial light and air pollution. Under dark skies the Trifid Nebula is stunning in an enhanced view, revealing most of its famous features even to a novice. It is sufficiently bright that by merely using conventional viewing methods it is not too challenging to make it out against the night sky. With conventional viewing and using an Orion Ultrablock filter with my 18 inch, I can usually detect color (pink and blue). Surprisingly, this is even the case from my suburban home site, far from an ideal location if ever there was one; unfortunately, it does not show at all in such conditions using the image intensifier. At our dark sky site, the sheer brilliance and easily discernible form of the Trifid becomes quite impressive and enhances any view we have now. Its celebrated

Figure 5.5. The Trifid Nebula M20.

form is especially clear; in the same conditions lesser apertures should provide quite a spectacle as well. The dark lanes, so significant a part of its character, almost leap out of the eyepiece or monitor view, dividing up the nebula into three primary segments, essentially meeting up near its center. No imagination is needed to fill in the blanks; this is the nebula of legend.

M20 is comprised of two distinct sections, one emission portion to the south (the larger, more famous portion with the dark lanes) and the other, primarily a reflection nebula, to the north. The fact that the emission component is so very greatly enhanced in the intensified view is not surprising in itself, but I was unprepared for the stunning account it gave of itself. I was also struck by the intensity in the intensified view of the central illuminating star, which is, in fact, comprised of no less than six stellar components. With sufficient magnification, which reduces the blinding glare, it will easily resolve into several components under intensified viewing. In such live views, the famous multiple shows clearer and even more prominently than by most imaging methods, which tend to fog out the region and obscure it. More striking yet in our enhanced view is the great delicacy of the dark lanes themselves, which have all the clarity and presence of time exposures. In my intensifier, the reflection portion to the north essentially disappears. Intensifiers normally produce a circular glow around bright stars; such is present in the image here at the center of the reflection component. Don't be fooled into thinking that this illusory artifact is, in fact, the reflection nebula! However, it does indeed provide a good simulation for the purposes of completing our expectations. StellaCam or the STV may possibly provide better views of the reflection component. All in all, M20 remains one of our foremost nebula subjects for any form of viewing.

Not unlike the Trifid Nebula is the much more difficult to see Cocoon Nebula, IC 5146, in Cygnus (Figure 5.6). The cluster within it, providing the energy by which it shines, is also included in the designation. Situated at one end of the dark nebula B168, the famous "cocoon" is presumably an illuminated portion of this dark stretch of gas and dust. Not particularly successfully observed with most forms of viewing, this otherwise faint nebula has become known to most people

Figure 5.6. The Cocoon Nebula IC 5146, by John E. Cordiale. (Courtesy Adirondack Video Astronomy.)

through the many beautiful color images that have been made with long exposures. I was therefore particularly impressed with the view of it obtained by the StellaCam II. It seems that such faint nebulae may well be this camera's strongest suit, giving us a view in the field utterly unknown to most amateurs in the past. I think you will agree it is little short of remarkable that we may finally see it while away from home with some degree of its magnificence, although, even in the best circumstances, it is reasonable to assume it will not reveal itself, live, quite as dazzlingly as in the frame reproduced here.

In our brief review of emission nebulae, we cannot leave the Sagittarius region of the sky – the very center of the galactic hub – without visiting the Omega Nebula (Figure 5.7). This marvelous sight, also sometimes known as the Swan Nebula, is another of the finest sights we have for live viewing. As one of the brightest emission nebulae available to the amateur observer, it is so prominent that it shows relatively well even from very unfavorable situations, such as conditions of great light pollution and haze.

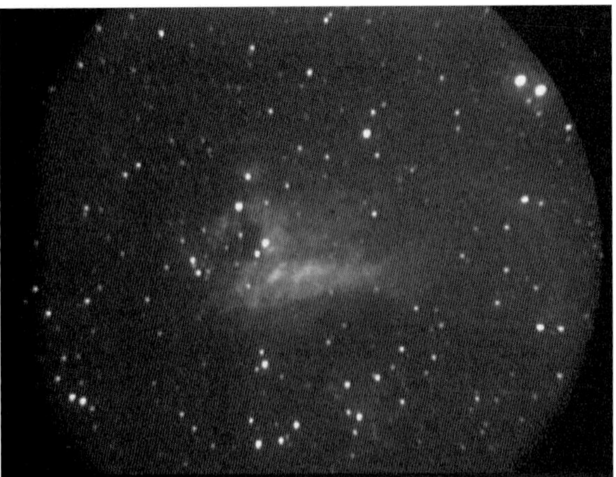

Figure 5.7. The Omega Nebula M17.

Although it has been variously described as appearing like the Greek letter Ω (omega), a comet, or even the number 2, it never strikes my fancy as appearing anything other than as a swan, riding gracefully through the "waterways" of the skies. The "waters" on which it floats so majestically are studded with glistening stellar points, like the sun sparkling on ripples in a lake or body of water. It is interesting to view the great bird with a conventional eyepiece before proceeding to enhancing devices. In this mode of viewing, you may be struck by how dark the region beneath the swan looks, again adding to the impression of floating upon dark waters rather than just empty space. This can only be due either to dark gases present, or more likely, contrast, due to the fact that the nebula spreads outward and much farther around the swan, and with more complexity than may be ascertained with live viewing. However, having seen long exposure images, I am aware how this dark "watery" area becomes well illuminated, so I cannot fully explain why the region looks so very dark in the conventional view.

Be sure to take in all the complexity of the swan's feathers; even a folded wing may be seen on its side. It is quite amazing how much refinement of detail can be made out, but not surprising when we realize that M17 is among the brightest nebulae known, and therefore among the best in accessibility to the amateur observer. Perhaps the greatest asset enhanced viewing provides with M17 is the ease of registering the complex and fine detail of this subject.

On the other end of the list of grand nebulous sights is little NGC 2467 in Puppis (Figure 5.8), which actually turns out to be quite an impressive emission nebula. Prominent in small instruments, and of approximately the same angular size as an average galaxy, it nevertheless requires sufficient aperture to begin to make out its most interesting features. In the enhanced view, those striking aspects assume greater contrast and distinction. As with many other emission nebulae, a hot young star cluster at its heart provides the necessary radiation for its luminescence.

In the view here, you will be aware of a line of stars that seems to bisect the nebula north to south, a striking and unusual feature, and continues well beyond its boundaries in both directions. You will also be struck by the multitudes of

Figure 5.8. NGC 2467.

stars all around and within the nebula. This larger star field also contains several small open clusters, unrelated to the nebula. Inside NGC 2467, you will be able to detect several members of its own illuminating cluster, as well as some fine, curious dark lanes. A small central "hole," as well as a finer dark lane running almost southeast to the center, should be evident, as well as a brighter strip parallel to the line of stars. I apologize for the many somewhat triangular stars in the video image here, more evident than in most of my images; I mention this even though you probably understand from the earlier discussion why it is so. In order to present the much dimmer nebula with similar impact to the live view, it became necessary to open the iris of the video adapter wider than would be correct for these stars on their own. The effect is only compounded by the great number of stars in the field.

Planetary Nebulae

Actually, planetary nebulae are a form of emission nebulae, but we need to deal with them separately. They are a unique form of emission nebulae, coming into existence at the end of many stars' lives, instead of at the beginning. These subjects perform very well in far less than optimal circumstances, including light-polluted skies, but they come still further into their own at a dark sky site. However, striking differences are more subtle than with many other deep space wonders. Fainter planetaries may be seen in optimal skies, but occasionally, a familiar one reveals itself anew in a sky of true darkness, and the larger, diffuse ones are more likely to be strikingly viewable instead of faintly showing.

The illumination of planetary nebulae, just as in all emission nebulae, is also indirectly supplied by stellar energy. However, in the case of planetaries, this is from one central star only, dying in slow motion as it sheds its mass outward. Reflected light alone of these stars would not be sufficient to render their surrounding gas shells visible, were it the only energy being radiated. Planetary nebulae, therefore, become visible instead because their central stars excite the ions in their surrounding gas shells, which then become visible as radiated light; they are true emission nebulae. The spectrums of many of these are rich in wavelengths particularly suited to image intensifiers and other infrared-sensitive enhancing devices. However, it also has to be said that brightest views tend to be with the smaller planetaries, regardless of our location, as these often have the greatest concentrated surface brightness; there is nothing surprising in this. Even so, some of what would appear to be the less-than-best-suited larger planetaries reveal components otherwise unseen, even when the gaseous shell may actually be harder to see.

Of the larger planetaries, perhaps the most celebrated is Ring Nebula M57 (Figure 5.9). Although a relatively extended object, it is actually a highly successful candidate for enhanced viewing from any surroundings. While the celebrated central star emits light predominantly in the blue part of the spectrum, it is usually easily visible through my 18 inch telescope with image intensification, even in the poor sky conditions most of us experience in the suburbs. It should be born in mind that this (approximately 16th magnitude) central star is often unseen with conventional viewing even through large professional instruments; I

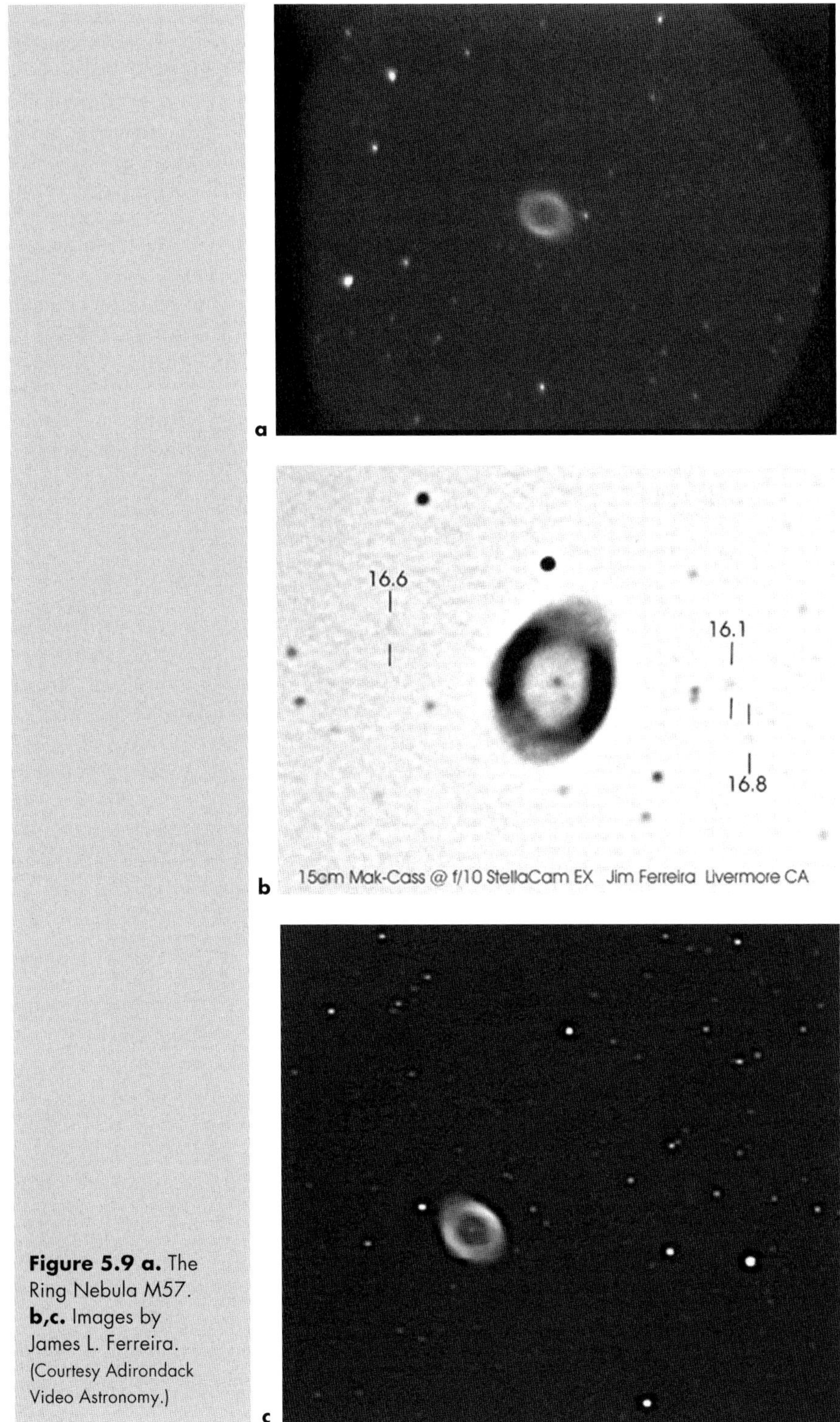

Figure 5.9 a. The Ring Nebula M57. **b,c.** Images by James L. Ferreira. (Courtesy Adirondack Video Astronomy.)

have read of one observer only uncertainly glimpsing it through the Mt. Wilson 60 inch (1.525 m) telescope. Admittedly, the conditions at the time were reported as not being great, but the 60 inch still has a gigantic aperture by any of the standards that amateurs normally use. Interestingly enough, the central star's fainter companion is also often easily seen through my intensifier, due to the specific spectral response of this device and the favorable wavelengths of light the star emits. In video images taken with the Astrovid 2000 camera, it seems to appear of equal or even greater brightness, for reasons I cannot explain, since the camera's own high infrared response cannot be a factor. (This is because it joins the optical path last, registering the green light only of the intensifier itself.)

Figures 5.9b and c will give you a very good idea of the potential of CCD video systems such as the Astrovid StellaCam EX, with these two additional images of M57. The smoothness of image in the compounded frames are readily apparent. Figure 5.9b (reversed as a negative image to show inner detail) utilized only 15 cm aperture, while Figure 5.9c was imaged through 25 cm aperture.

At a dark sky site the entire structure of M57 is a dazzling showpiece, with all kinds of subtleties jumping into view; the central star is now a very easy mark in only moderate amateur apertures. Surrounding the nebula are all manner of faint stars usually seen only on time exposures, which complete the view. The legendary nebula appears much like a bright and freshly puffed smoke ring, wafting out into space. With direct viewing through the image intensifier, you will notice many subtleties of shading and texture in the ring itself, which do not translate to the recorded view, or even that on a monitor. Even easier to see (and record) are the variations in brightness of the ring, which appears brightest along the longer sides of the oval-shaped formation. The central hollow is clearly lighter than the empty space surrounding the nebula itself, but I have yet been unable to detect the striped bands sometimes reported by observers with large telescopes, and which so clearly show on many photographs.

The Dumbbell Nebula M27 in Vulpecula is one of the biggest and brightest planetaries in the sky (Figure 5.10a). It is easily visible with small apertures, and even with binoculars. While it is true that the view of it is quite impressive with a narrowband filter, even in the city, once we turn our electronic equipment on it, it may be a different matter entirely. From such places as my own suburban location, I can hardly see the nebula shell at all through my image intensifier. However, it did provide a somewhat different perspective of the nebula which I was unable to see with the filter, that of the inner gas bubble and the many stars framed by it. Now, under dark skies, it had a surprise in store.

The famous round gas sphere was not only readily apparent, but quite bright in the intensified field of view, with the central star and all manner of tiny stars suddenly coming into visibility. In fact, the sight immediately reminded me of many of the famous astrophotos I've seen over the years, reminiscent of a partly eaten apple. Within the brighter portion, you will be able to make out interesting variations of brightness, as well as numerous tiny stars, especially the central illuminating star itself. You may notice how the visible central cross-shaped structure seems to be influenced by the presence and placement of these stars, something I commented upon in my previous book *Visual Astronomy in the Suburbs*.

However, there are quite significant differences between viewing M27 this way and viewing conventionally with a narrowband filter. Such a view, by compari-

Figure 5.10a. The Dumbbell Nebula M27.

son, shows another face. The round sphere is now contained within a larger, oval whole, and the central star is hard to find. I am including my drawing of the nebula, as seen from the suburbs with light filter. This image was reproduced in *Visual Astronomy in the Suburbs*, and I think its inclusion here is instructive in that it illustrates these two important structural facets of the whole.

You will see from the drawing that M27 is, in fact, a rather egg-shaped object when we are viewing the entirety of its form (Figure 5.10b); the video image, taken under dark skies, reveals a different manifestation as seen in intensified viewing. The drawing shows the central portion of the whole appearing broader, and M27 looks more like a traditional planetary form, an encircling shell of gas. In such conventional views, it suggests to me a large soap bubble, gradually becoming unevenly distributed, floating precariously through the air, waiting to burst at any moment. By comparison, the video image shows only the brighter part of the bubble: that mottled, uneven portion that gives rise to its name, the "Dumbbell." The "missing" sides can vaguely be seen, but need more to be imagined in your mind's eye.

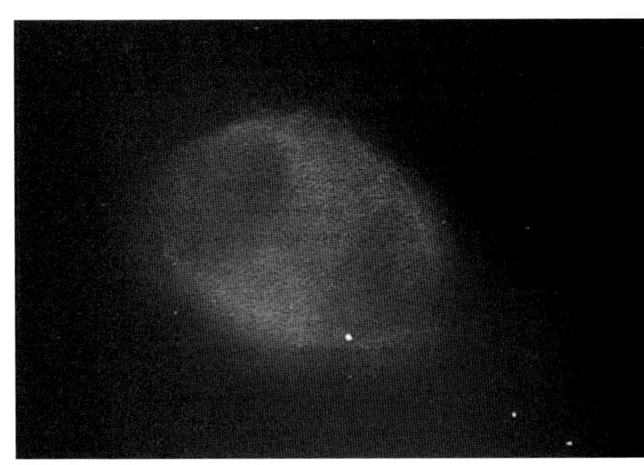

Figure 5.10b. Drawing M27.

Here also are processed images of M27 made with two apertures and two generations of StellaCam cameras. Again different from the conventional view, they are displayed with ever-increasing advantages of aperture and camera. Figure 5.10c was imaged with a StellaCam Ex camera through only 15 cm aperture. Figure 5.10d was through 25 cm. However, Figure 5.10e was produced not only with 25 cm aperture, but also the new StellaCam II camera, showing the increasing resolution and exquisite subtleties possible.

Often a source of disappointment to the amateur observer because of its relatively small dimensions, the Little Dumbbell Nebula M76 is nevertheless one of the more prominent planetaries we may view (Figure 5.11). It becomes partic-

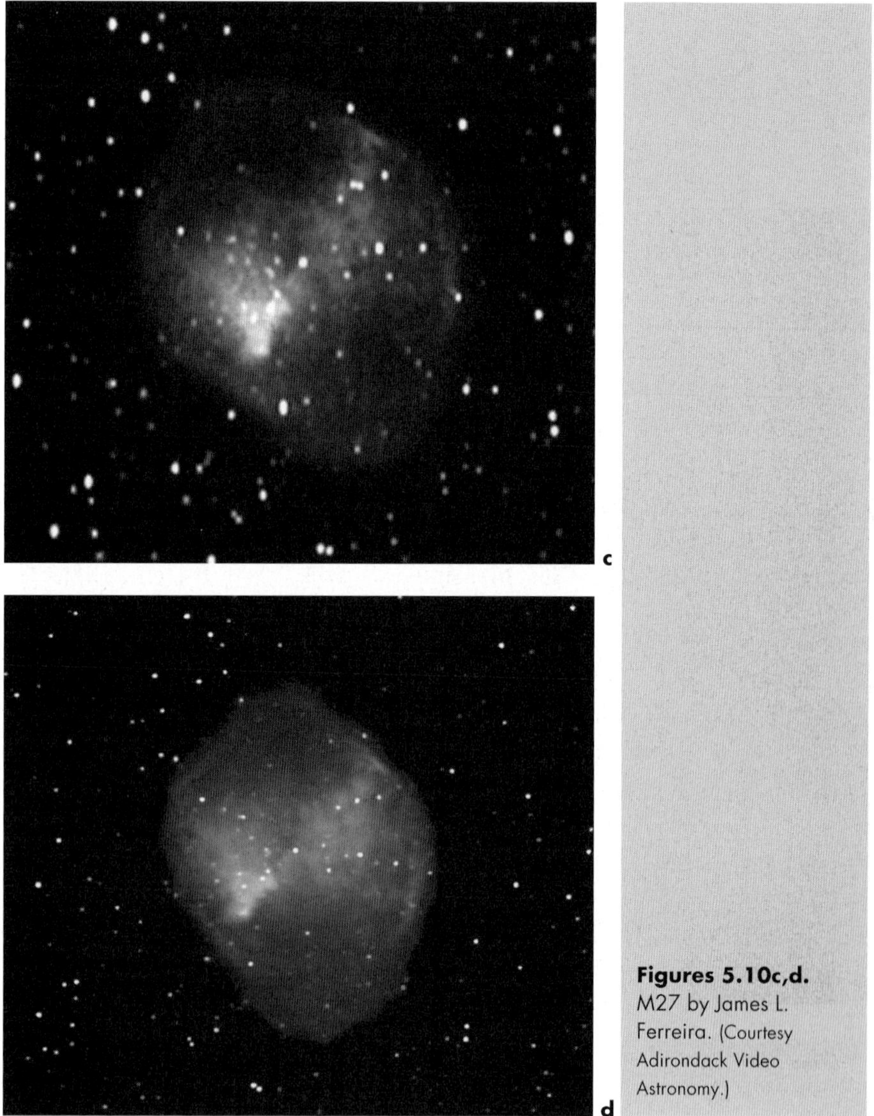

Figures 5.10c,d. M27 by James L. Ferreira. (Courtesy Adirondack Video Astronomy.)

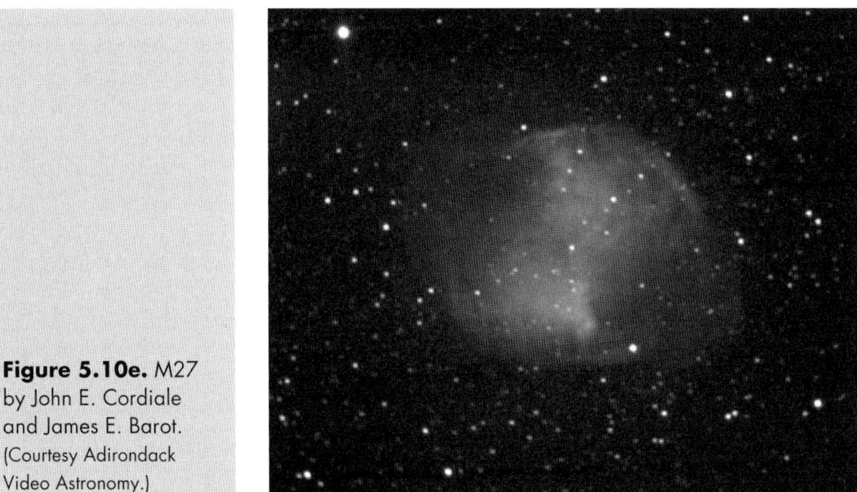

Figure 5.10e. M27 by John E. Cordiale and James E. Barot. (Courtesy Adirondack Video Astronomy.)

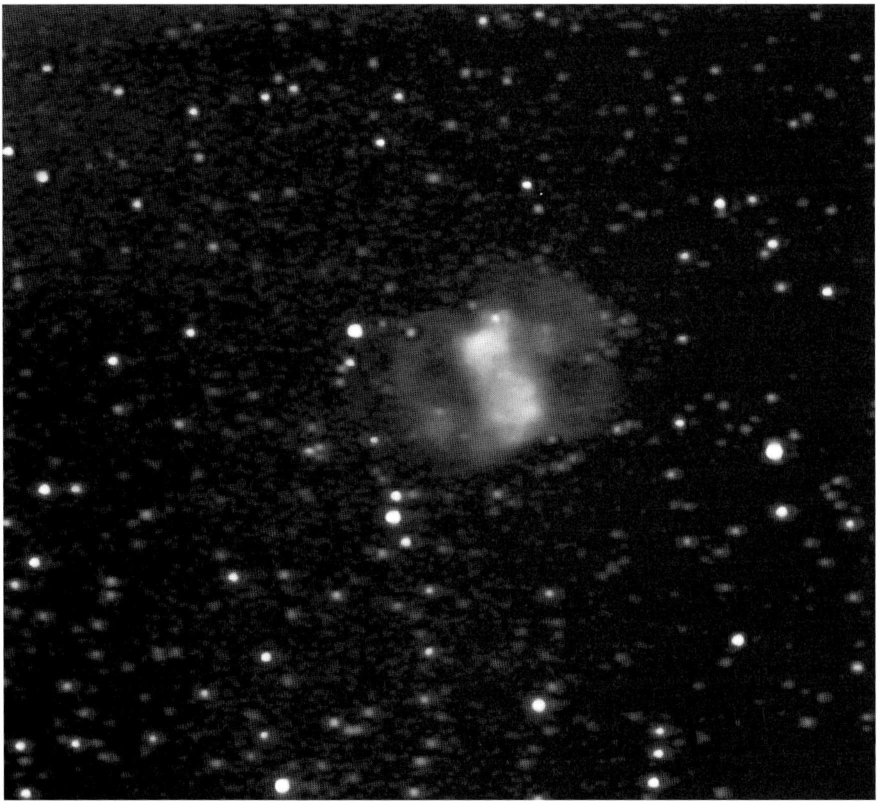

Figure 5.11. The Little Dumbbell Nebula M76, by James L. Ferreira. (Courtesy Adirondack Video Astronomy.)

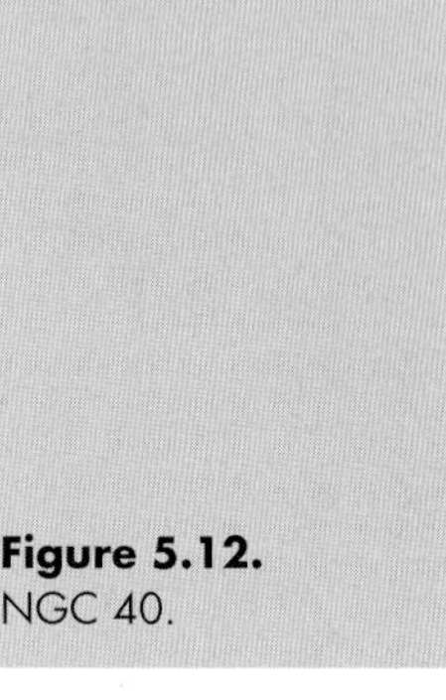

Figure 5.12. NGC 40.

ularly interesting under the true cover of darkness found at remote sites because the butterfly-like "wings," those extensions curling around its extremities and normally unseen in conventional viewing, provide new potential to us with our enhancing devices. The view here, imaged with a StellaCam II, represents probably the ultimate video image possible today with most amateur telescopes.

NGC 40 in Cepheus presents another amazing case in point for the benefits of enhanced viewing (Figure 5.12). Relatively small and decidedly unexceptional in the suburbs with conventional viewing, it suddenly reveals its true form when seen with intensifier, even in these challenging surroundings. In true dark skies, the sight is further expanded, as the surrounding gaseous shell is thrown into greater contrast against the night sky. Typical observers' descriptions with conventional, possibly filtered viewing describe a disk with a prominent central star (actually, quite easily seen even from the city without enhanced viewing). Usually some comment is made about brighter portions of the surrounding ring, along with some annularity visible with larger apertures, but certainly no more than this.

With enhanced viewing, it is newly revealed to us as a truly wondrous object. The 12th magnitude central star is ablaze in the view. More striking, though, is the extreme annular effect of the surrounding disk itself, and more than just the original circular ring jumps out at first glance: an additional oval ring is now visible, although in some moderate apertures, an incompletely resolved outline may show it appearing more like spiral "antennae." The lesser axis of this ring coincides with the brighter segments of the inner ring, which may, in fact, be the reason for the observed brighter portions of the inner ring in the conventional view. Truly, when viewed in this manner, this nebula would have deserved the title "Saturn Nebula" much more than its better-known actual namesake, NGC 7009.

NGC 6826 in Cygnus is actually quite typical of the way most moderately bright planetary nebulae of smaller dimensions appear in enhanced viewing (Figure 5.13). Regardless of the method used, the results are fairly similar, and often there is not a huge difference from that obtainable by the same means in quite light-polluted skies. This celebrated sight is often known as the "Blinking Nebula" because of its tendency in conventional viewing to reveal either the central star or

Figure 5.13. NGC 6826. StellaCam EX image by James L. Ferreira. (Courtesy Adirondack Video Astronomy.)

alternately the nebula, depending on how one looks at it. In enhanced viewing, however, all of this is moot, and the little nebula shows not only the central star and gas shell simultaneously, but also an inner gaseous ellipse, much like the Eye Nebula, NGC 3242, sometimes also known as the "Ghost of Jupiter." I include this image here as a reference, but by all means, don't take any potential spectacle for granted; you never know just what will show up better than the expectation. Certainly, dark skies open many new doors.

For purposes of enhanced viewing, the grand Helix Nebula is a nonstarter in situations where light pollution is a factor (Figure 5.14). In fact, this planetary is actually a difficult real-time object in any circumstances, and has become known to us more through time exposures than detailed descriptions through the live view, enhanced or conventional. Even though still not exactly startling under

Figure 5.14. The Helix Nebula NGC 7293.

dark skies, and nowhere close to the magnitude of the Dumbbell or Ring Nebulae, with enhanced viewing it does at least allow us to look at it a little more easily. It readily shows up as something of a complete and defined circle of gas, complete with central illuminating star. Additionally, and in many ways more interesting to me, are the numerous tiny "inner" stars that now become easily visible within this gaseous disk. Including the central star, I can usually make out at least 12. Brighter portions of the surrounding gas ring are visible to the east and west, and even a hint of the helical twist, although I caution you, this is very difficult to detect. Nevertheless, it is there, and patience hopefully will reward your efforts. I find it very satisfying to be able to make out more than the usual faint blob of light in the field of view.

Extended Nebulae

For the purposes of this writing, we will characterize those nebulae that are so vast as to defy normal telescopic viewing as extended nebulae. For example, if we could view the entire regions of nebulosity encompassing the Lagoon Nebula and the Trifid Nebula, or the region surrounding the Great Nebula in Orion, we would certainly think that they qualify as extended nebulae by this definition, as these famous landmarks are only part of far grander nebulae. The only problem in so characterizing these examples, of course, is their invisibility as extended structures by any type of live viewing.

As you know, pure emission nebulae are likely to make very impressive enhanced telescopic viewing, as long as they are bright and concentrated enough. The North American Nebula NGC 7000 qualifies as both an emission and an extended nebula (Figure 5.15).

The full extent and shape of the North American Nebula NGC 7000 in Cygnus is far too large and diffuse for normal telescopic purposes. Because it is quite faint and widespread, it is normally considered a photographic or binocular object only. However, the entire nebulous region shows up quite well with no magnification (the ultimate wide field view!) utilizing the 50 mm F1.3 Collins

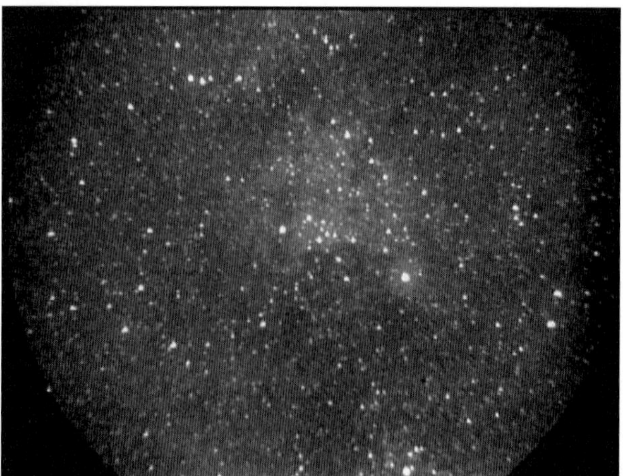

Figure 5.15. The North American Nebula NGC 7000.

primary lens attachment on the I_3 Image Intensifier. Because the red spectrums of nebulae such as this generate their own light and are excited into fluorescence by ionization instead, NGC 7000 is well suited to Generation III intensification, and is reasonably conspicuous in the field of view.

Taking its name from a rather remarkable attribute of its outline, the North American Nebula is almost an exact map of the continent in shape. Utilizing a wide-angle lens such as the one described above, it will show us something of its form to us, live, with our enhancing devices. The "Gulf of Mexico" may be seen at lower middle; nearby, in the "Atlantic Ocean" lies the very faint Pelican Nebula IC 5067-70, difficult to make out here. Because of its specific spectral makeup, you will see a somewhat different face of NGC 7000; look carefully, and its familiar form will begin to make itself known. Many other dark nebulae (see the next subsection) exist in this region, as they do throughout the girdle of the Milky Way. In viewing extended structures in this way, we are now concentrating all of the tenuous and vague light available into a small field of view. The method provides an interesting new method of study and review for many other large deep space objects, not the least of which are dark nebulae, often vast structures in their own right.

Dark Nebulae

Those structures usually designated thus involve the vast unlit quantities of dust and gas circling around the galaxy. Most are interconnected and seen collectively as a dark band dividing the Milky Way in the edge-on view of the galaxy from our perspective, much like, say, that seen in the edge-on galaxy NGC 4565. The legendary nineteenth- and early twentieth-century astronomer E.E. Barnard assembled a large catalog of dark nebulae, bearing his initial (B) in their designation. However, not all dark clouds precisely fit the general stereotype, and may belong primarily to smaller structures. From a standpoint of visibility, the most impressive of all may be situated placed in front of emission nebulae or star fields, and some are incredibly impressive.

The Eagle Nebula M16 in Serpens, in addition to the sight of the showy star cluster within it, is one of the showcases of the sky. Regularly a photographic favorite of astronomical publications, it is by these that we have come to be so familiar with this spectacle. The strong visual suggestions in M16 have made it one of the most recognized of all deep space objects. This nebulous region consists of both emission and dark nebulous matter, and certainly this would suggest that it has potential for us with image enhancement. However, in live viewing it always remains a challenge, no matter what we do. We should only expect it to be modestly bright at best, even with larger apertures, as the illuminated emission part of the nebula is not especially radiant. Many observers of antiquity were actually completely unaware of the nebulosity, even more the "eagle" itself, a classic dark gaseous formation superimposed on the whole. The famed observer of antiquity, Admiral Smythe, only commented upon the open star cluster within it, M16, and certainly even Messier himself had no awareness of the nebula's existence. (See Figure 5.16.)

The dark portions of M16 are actually difficult features to discern easily in any form of live viewing. We have become so accustomed to the well-known portraits

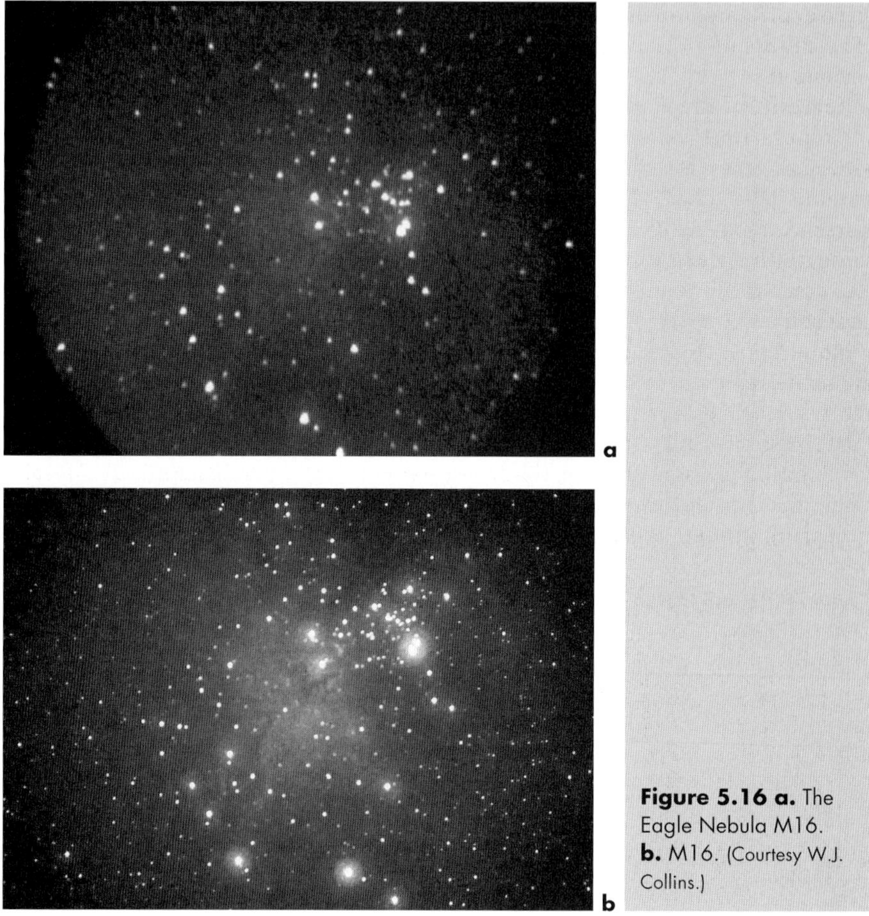

Figure 5.16 a. The Eagle Nebula M16. **b.** M16. (Courtesy W.J. Collins.)

of M16 that we may be disappointed even with the live enhanced view, but patience and experience will serve you well. With time, careful adjustment to the view will gradually condition the eye and mind to discern the features it is famous for. The large wing in flight is visible, even the outstretched talons underneath the eagle itself can just be made out, as well as traces of the trail in the eagle's wake. With conventional viewing, I believe the novice will have little chance of making out anything of its outline, but may fare much better with intensified or other suitably enhanced approaches. It is quite a significant prospect, and offers us such a dramatic image that it will be worth the considerable perseverance it requires.

In the video frame (left, Figure 5.16a), we must therefore make allowances for the less than optimal brightness, and particularly the fainter than desired eagle component. However, if you look carefully at this image, and (most importantly here), together with a good reference image, such as the 6 second intensified digital exposure at right, made with just 7 inches (178 mm) of aperture (Fig 4.43b), you will see all the primary ingredients of what makes up this splendid sight. The celebrated "eagle" that we all look for looms upward out of the cloud. This dark formation has the remarkable capacity to suggest not only a great eagle soaring in flight, but the very motion of flight itself. The Hubble Telescope view of these exploding projections of hot gas, better known to us as by its famous title "The Pillars of Creation," further revealed the vast processes involved in star formation.

Regretfully, the famous (and often infamous) Horsehead Nebula in Orion is not a prime target, even armed as we are with our powerful light enhancer. Although it is set in front of an emission nebula, IC434, this particular background is so faint that it does not provide sufficient contrast against the Horsehead to benefit us sufficiently. However, there is an additional problem that comes with a warning! The bright star, Zeta, is so close to the Horsehead itself that you would be best advised to steer well clear of this potentially damaging light source as a user of an image intensifier. Such a bright light is potentially as ruinous to an intensifier tube as is a direct, especially a telescopic view of the sun to the human eye. In monitor viewing, it may be outside the oblong field of view but nevertheless present on the intensifier tube.

In real-time viewing, one of the best revealed of all smaller dark nebulae is the Ink Spot B86, lying within the Great Sagittarius Star Cloud (Figure 5.17). Though small, the Ink Spot becomes nothing less than stunning in the enhanced view, appearing shaped like an anvil, with a bright star almost touching its broadest side. The effect of contrast, seen live, is far greater than can be shown in the single frame reproduced on the page; to me, it is among the most starkly black of all such nebulae. Not just confined to its most immediately apparent outline, there is an additional streak of dark material shooting out SW toward the nearby open star cluster, NGC 6520, which only compounds its dramatic visual impact.

There are countless other small dark nebulae scattered throughout the band of the Milky Way, and maybe you will be inspired to explore them. Many of these dense, independent clouds of dust and gas are true Bok Globules, which may ultimately collapse to become stars. Be sure to try your luck with such notable examples as B143 (Cygnus), B62, B63, B68 and the dark extended fingers of B276 (Ophiucus), B90 and B92 (Sagittarius), B50, B53, B253 (Scorpius), the triangular B104, and B312 (Scutum); the list goes on. You can find a complete list of Barnard's dark nebulae at the Web site of the Belmont Society, www.belmontnc. 4dw.net. (This site also contains much additional information for the amateur astronomer.)

Figure 5.17. The Ink Spot B86.

The many grand-scale, extended dark nebulae of the Milky Way practically become a separate category in themselves. To the uninitiated observer, it might appear instead that there exist great voids within the stellar fabric. Appearing as huge dark gaps within the vast star cloud regions, it is as if the stars behind them suddenly have been covered in a blanket. They are, of course, mostly connected to each other, existing as part of the Milky Way's great dark rift. Viewing them can be a problem, since the somewhat limited field of view of our telescopic equipment will not allow a sufficient portion to be contained within the confines the devices allow us. Our best chances to see them telescopically may be when they are small enough that major portions are able to fit within the somewhat restricted fields of view of most telescopes, as well as being set against concentrated fields of stars. Otherwise, the 1X primary lens attachment will serve us admirably once again, and provides an ideal answer for extended examples.

Around the very core of our own Milky Way galaxy, especially in the Sagittarius and surrounding regions, lie some of the best examples of dark nebulae, lying tangled among great fields of stars. Other breathtaking star fields of the Milky Way can be found scattered across its great girth as well, of course, but it is especially around the highly luminous center that many of the greatest prizes of the dark nebulae are set in starkest contrast against it. In these seeming voids exists a staggering amount of interstellar gas and dust. Unlike some smaller bright nebulae on which dark nebulae are superimposed, most of the crowded star fields on which the grander dark nebulae are projected are not designated as specific deep space objects; we are witnessing the very structural fabric of the galaxy itself. Many dark nebulae appear as branches of twisted dark blotches, branches and strands. Suffice it to say, Barnard's catalog of individual numerical listings frequently overlap into each other. It is quite typical for several numbers to make up one major dark nebulous region, so unless you are particularly interested in making a detailed study of dark nebulae, it would serve little purpose here to dissect the views reproduced here in this manner.

The majority of the most striking examples of dark nebulae lie closest to the center of the galaxy, around Sagittarius and Scutum, because of the greater concentrations of stars or other luminescence upon which they are projected. A superb example for our viewing of extended nebula is the Pipe Nebula, actually a combination of several cataloged dark nebulae (Figure 5.18). Not that this should matter to us, it still appears as one: B78, with B59/65/66/67, is probably the best-known named large dark nebula. Comprised of the typical interstellar gas and dust that forms the visible belts of other galaxies when seen edge-on, it can be well seen at very low powers. Peering into the Milky Way edge-on, the significance of the dark nebulae for scientific study is never more apparent, especially as it relates to other galaxies.

The Pipe Nebula, seen in the image at the upper left, is situated very near the heart of the galaxy from our perspective in space. The stem and bowl are quite clear. No less striking are the other dark nebulae all over the field of view. The region of the hub of the galaxy itself is, in my view, the most visually stunning real-time view of all star fields with dark nebulae, and is the brightest portion, better known as the Sagittarius Great Star Cloud, one half of the galaxy's central core. The actual heart of the galaxy, however, lies within the dark nebula between it and the pipe bowl. In visual impact, at least in live viewing, especially and sur-

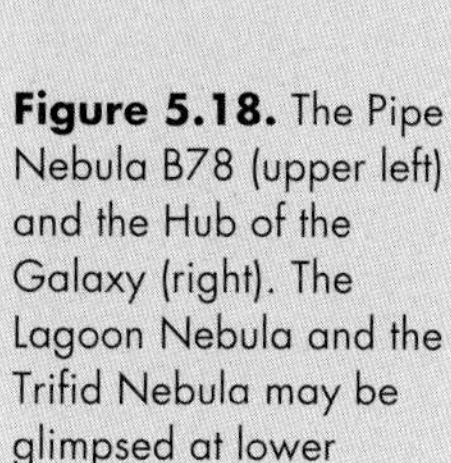

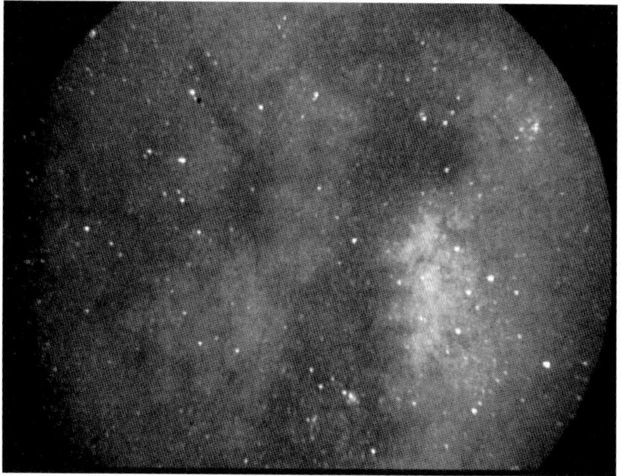

Figure 5.18. The Pipe Nebula B78 (upper left) and the Hub of the Galaxy (right). The Lagoon Nebula and the Trifid Nebula may be glimpsed at lower middle.

prisingly on a monitor, these views certainly equal anything I have ever seen in time exposures.

At the NW edge of the pipe bowl is an additional astounding sight that requires viewing through a telescope because of its small dimensions. The Snake Nebula B72 is always difficult to view live, but the effort to do so will pay off with dark transparent skies, and it is even more readily seen with image enhancement (Figure 5.19). The video frame reproduced here shows it faintly, and it will be more apparent with indirect vision, where it will be seen winding its way across the field.

You will surely want to explore the realms of many other dark nebulae throughout the great band of the Milky Way. They present us a whole new range of viewing possibilities under the canopy of a truly dark sky, the scenes hard to describe in earthly terms. Indeed, at a dark sky site, the billions of stars of the Milky Way glow like iridescent artwork, painted in great swaths across the heavens. If these spectacles don't inspire you, nothing will!

Figure 5.19. The Snake Nebula B72.

Of course, imaging with time exposures is traditionally the most impressive method for gaining access to these wonders, but we cannot see the results in real time. Up until now, if we wanted to see them live, our choices have been to use binoculars and richest field telescopes. Great as this traditional viewing method is, it is nevertheless an entirely different viewing experience from the new systems available. Nothing quite prepares you for the luminous, dazzling spectacles such as a 1X power viewing lens attached to an image intensifier can provide! Merely hold the device in your hand, just as you would a monocular, and you will be freely able to scan wide regions of the sky to great effect. Remember of course, that at really low powers, or no power at all, everything emitting light appears exaggerated in the field of view. Therefore, the darker the sky, the better the result will be, because we want the background to appear as contrasted as possible and not overwhelming the scene with a green wash of light.

The views are also quite astounding on a video monitor, their luminescence almost leaping from the screen; you will probably be amazed by the brilliance and revelations of all that you can see. The integrating CCD video systems I have described will not allow such a direct use as panning, of course, as when hand-held they will not permit frame compounding or extended exposures. By no means a problem, we can simply attach the camera's components to a tracking telescope or mount. In fact, I usually attach my Collins intensifier system itself directly to my telescope, taking advantage of the stability this provides. This becomes necessary anyway when I want to record images or view them on a monitor. Many large and prominent dark regions may easily be discerned all around the Milky Way this way, and good results with our own local star fields are among the easiest and most exciting forms of viewing and imaging that we have available to us. There is hardly a view that will not reveal something of interest.

Figures 5.20a and b are good examples of dark nebulae and Milky Way star fields in Sagittarius/Scutum regions. Figures 5.20c, d, and e are some examples of dark nebulae within star fields of Cygnus.

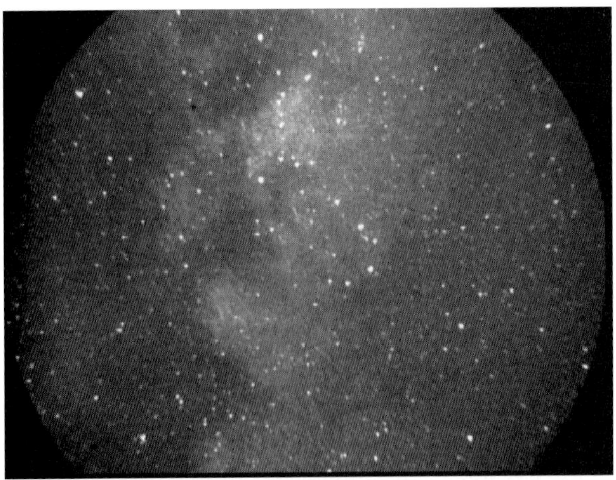

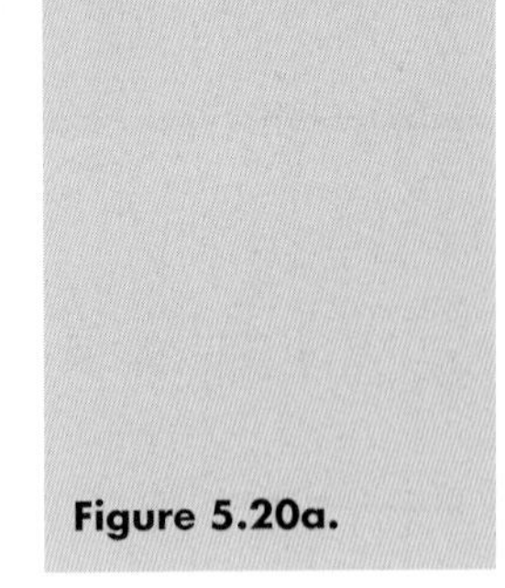

Figure 5.20a.

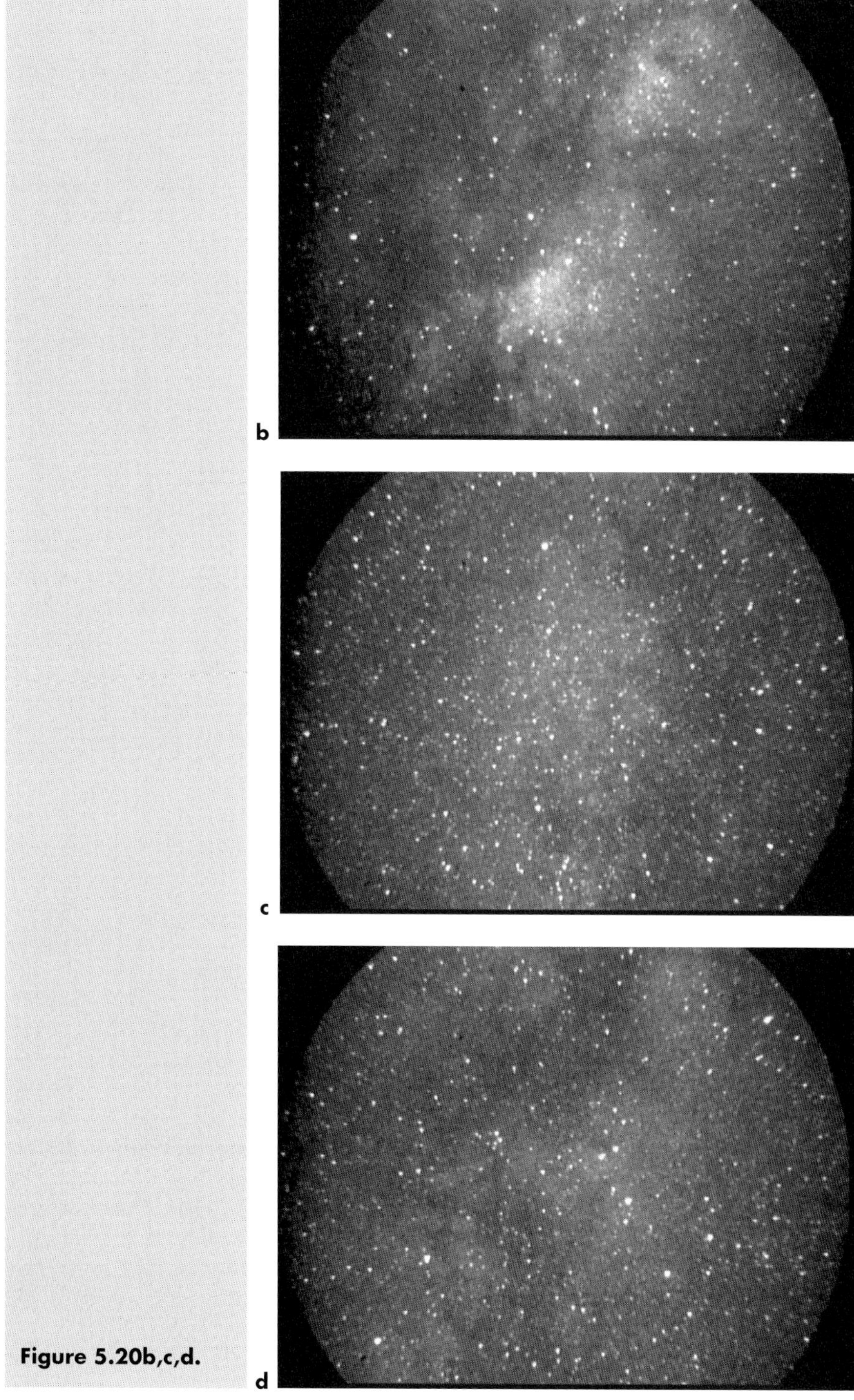

Figure 5.20b,c,d.

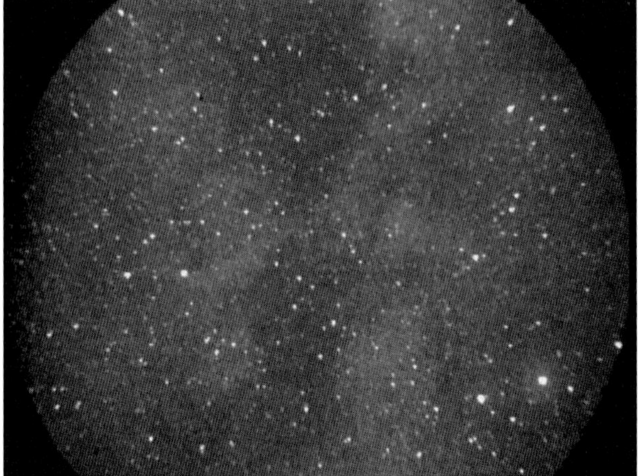

Figure 5.20e.

Unusual Nebulae

We also have those nebula that exist by extraordinary circumstances. Hubble's Variable Nebula NGC 2261 in Monoceros is an interesting study from any site, but all the more so under dark skies (Figure 5.21a). I have always had pretty good results with this subject even from the suburbs, but in true darkness, it allows us to discern noticeably more of the inner delicate features for which it has become so famous, because of the greater contrast and brightness provided by the better location.

Hard to categorize, the nebula is probably just a portion of a larger whole, which happens to become illuminated by localized ionization emanating from the hazy star at its tip (made so by dusty obscurities?). However, the well-known dark shadows within the nebula itself, which change shape and number over short periods, would seem to contradict any theory of the nebula producing its own light. Nevertheless, it is only logical to assume that it has a significant portion of emission component in its makeup because it is so prominent under image intensification. (A pure reflection nebula would tend to disappear in the intensified view.) The shadows arguably could only be caused by obscuring matter swirling close to the illuminating star, as the scale of the nebula is too great to allow such true physical changes at what would amount to speeds close to that of light itself. However, such cause and effect should have no effect on an emission nebula's luminosity! So it would seem Hubble's nebula must contain elements of both reflection as well as emission. Presumably, the shadows show partly because of relative contrast with the fully illuminated portions of the gas cloud. These varying dark shapes, often invisible even in quite sizable scopes, make it across the visibility threshold quite readily with my image intensifier. They certainly intrigued Hubble for years. Completing this ghostly sight is the semicircular setting of stars framing it against the void. (See Figure 5.21b.)

One of the most extraordinary nebulae known to us in the Milky Way is the cosmic remnant of the colossal stellar explosion in AD 1054 – a supernova – so

Figure 5.21
a. Hubble's Variable Nebula NGC 2261.
b. Hubble's Variable Nebula, surrounded by its semicircular halo of stars.

bright that it lit up the daytime sky, leaving us with its tattered form, and better known as the Crab Nebula M1 in Taurus (Figure 5.22). Still expanding at an easily measurable rate from year to year, the Crab offers some visual opportunities for the amateur observer with conventional filtered viewing, and also with image enhancement. Usually difficult or even impossible from suburban locations, it is at once easy to see from a dark sky site. It appears almost bright, a special treat compared with its vague visual nature in the city. (Because it is basically a nebula of the emission variety, one would think that it would always lend itself to image enhancement, not just in dark sky country.) Interesting views of it may also be had with conventional viewing and an Orion "Ultrablock" filter, where the many complex tendrils we associate with this particular object can become visible on transparent nights, sometimes even within city limits.

I should say that some fairly impressive (and bright!) views of it can be obtained with enhancement at dark sky sites, and at times I have even sighted the

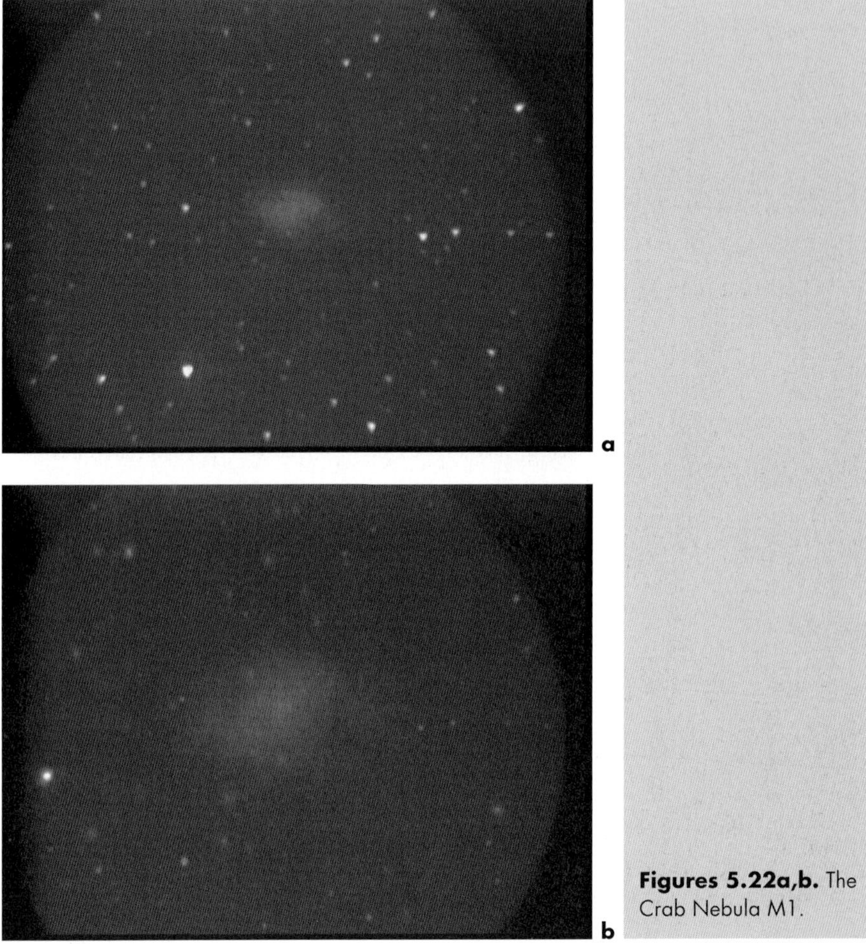

Figures 5.22a,b. The Crab Nebula M1.

16th magnitude neutron star at the center of this tangled star wreck (Figures 5.22c and d). This is the greatest bonus of all. We should not lose sight of the fact that this star is no more than a few miles in diameter, and here we are, viewing it across unimaginable distances. I have not read of amateurs having sighted this star with any given telescope aperture in any conditions in conventional observing. However, I have even glimpsed it (and imaged it as well) through my Collins I_3, from my home location in the greater metropolitan Los Angeles area. I believe this to be remarkable and worthy of special mention here, particularly since we are looking for extraordinary sights. The four images reproduced here show the nebula and some detail quite well. The second, a larger-scale image (a separate exposure on a steady night), shows more clearly many fainter stars, and the famous neutron star just makes its presence felt. I have seen photographs, exposed without image intensification for up to an hour through similar-sized telescopes, that barely reveal more. The third and fourth processed, compounded images taken with a StellaCam EX, aside from being presented very brightly here,

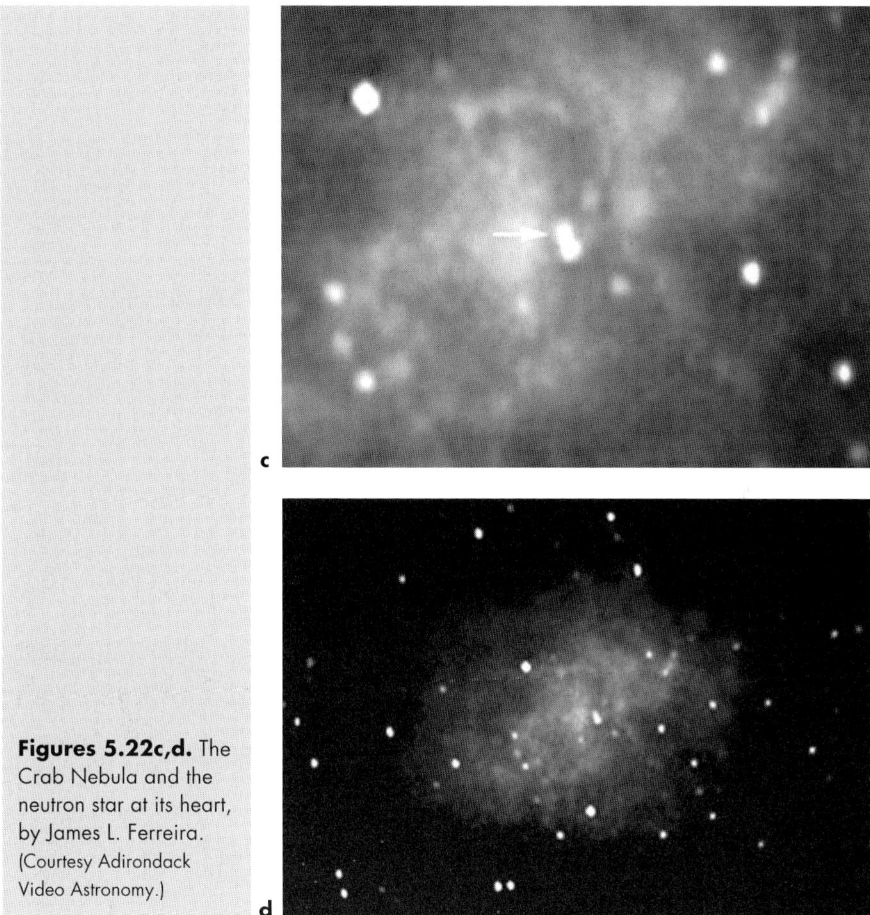

Figures 5.22c,d. The Crab Nebula and the neutron star at its heart, by James L. Ferreira. (Courtesy Adirondack Video Astronomy.)

also reveal the neutron star; the arrow pointing to it should make its position clear in all the images.

In conclusion, perhaps the greatest surprise would be the very subtle Veil Nebula NGC 6960/6992 in Cygnus (Figure 5.23). Long renowned for the very lack of success many observers experience with it in visual astronomy, the "Veil" finally began to come into its own with the advent of narrowband light filters. I have seen it to considerable satisfaction from the suburbs in the small hours of the morning with such a filter, although I cannot say it was better than faint at best. I say "considerable," because just a few years ago this would have been completely impossible!

To actually see the famous stellar fragments, live, in something approaching photographic detail is not anything one would expect in any circumstances. Although even at dark sites it remains always challenging with image intensification, it can reveal traces of its form with sufficient aperture. However, the NGC 6960 portion provided a stunning view in the new StellaCam II, the image here presumably being the result of maximum frame compounding and some

Figure 5.23. The Veil Nebula NGC 6960, by John E. Cordiale. (Courtesy Adirondack Video Astronomy.)

careful processing. Certainly this new CCD video device seems well suited to such subtle and faint celestial objects, and this is the first time, to my knowledge, that the Veil has been imaged satisfactorily in something resembling real-time.

Star Clusters

These great conglomerations of stars, amounting to well over a thousand within our own galaxy, provide some of the best opportunities for image intensification. These seemingly tightly packed stellar populations wouldn't seem so dense to a resident of a hypothetical planet circling one of their stars, since they are still far apart in true galactic terms. Most clusters belonging to the Milky Way have stellar populations ranging from under a hundred in open clusters, to as many as several hundred thousand and up in the grander globular variety. In many cases, they are slowly but surely being torn apart by the strong tidal forces of our own galactic system. Some may have even been captured from outside the galaxy; Omega Centauri NGC 5139, that unbelievably spectacular globular cluster, may well be the core remnants of a dwarf galaxy that strayed just a little too close to our own galaxy for survival. Unfortunately, for most Northern Hemisphere observers this greatest cluster of them all will be below the horizon; sadly, the same is true of its near rival 47 Tucanae. Happily, there are many others that will still blow the imaginations of these unfortunate souls away!

Those globular clusters that belong to the Milky Way, numbering about 150 in all, offer the greatest potential of all for us in our local neighborhood with enhancing devices. This is because they consist of relatively tight balls of lighted points, visually locked in step together. It is these clusters that have the greatest numbers of stars. By comparison, open clusters, far more numerous, are usually quite spread out, often appearing at their finest in binoculars or very low power/wide aperture "richest field" telescopes. The maximum number of stars comprising them is likely to remain under five hundred, and is sometimes very much less. Some of these clusters can be observed in the process of creation, forming from the gaseous matter of the great galactic nebulae of the Milky Way. Within the Rosette Nebula (NGC 2237) is a good example of the emergence of new stars forming in groups, as an open cluster. Typically, the fields of view of moderate or larger scopes are likely to be too small to contain the full dimensions of most of the best-known larger open clusters.

The most compact star clusters, of any type, are outstanding subjects for almost any telescope. The brightest among them provide our best potential for spectacular live viewing, although obviously, the larger the aperture, the greater will be the range of illumination and stellar resolution. Even conventional viewing can hardly be discouraged; indeed, this most traditional of all methods provides an exquisite and unique perspective, unmatched by any other, including the best that photography or the most revealing CCD image has to offer. While we can gain greater brilliance and resolution by other methods, the delicate, refined, almost three-dimensional quality that many of these clusters show in the conventional view is not easily forgotten. The qualities of stellar color, and that other intangible part of conventional viewing, which seems to make the clusters appear suspended in space, is regrettably absent in their new manifestation by enhanced viewing. Some of the closer and larger examples will knock even the most skeptical viewer's eyes out of their sockets – and that is long before we turn to the intensifier or CCD video!

Great additional visual insights may be gained from enhanced viewing. The degree of this depends on many things, including the specific spectrum of light reaching us, the range of which can be quite wide. This spectrum may be a consistent spectral type from the formation as a whole, or may occupy a wide range from different star types within the cluster. The exact makeup is at least partly determined by the specific cluster's age and type, and will become one of the most significant factors in how brilliant it is under intensification or other enhancement. The most strongly favored wavelengths are those emitted by yellow and red older stars, and clusters rich as these usually yield the best results.

Regardless of the particular spectrum of light being radiated by any given star within the cluster, stars are not diffuse or vague light sources. This means they all will benefit from the application of enhancing devices in general. Either a star can be seen as a point, or it remains below the threshold of visibility. Therefore, regardless of the type of light being emitted, the majority of clusters of all varieties in the spectrum will put on a pretty grand performance. With enhanced viewing devices, the differences in overall brilliance from one to another may be noticeable. Clusters of any given magnitude will generally appear more striking than other comparably bright deep space objects. This is precisely why they can be among the most successful objects for us to show to others; it does not take great observing skills to appreciate seeing them in the enhanced view. While the best are spectacular indeed, some of the lesser clusters, usually reserved for time exposures or mammoth telescopes, may be newly and astoundingly revealed. Even with moderate apertures, often formerly faint stellar points, or unresolved glowing regions, become present as resolved stars in the image. Many examples of dense cluster can be cracked wide open, or at least made to show a good part of their makeup. As long as stellar points are present and we are able to amplify them to a sufficient degree to see them as such, our potential is only limited by the specific cluster itself. This includes some of those usually considered too remote as to allow much, if any, resolution.

Although some of the qualities that are so magical in the conventional view are lost, the payoff instead is the sheer brightness and resolution that is possible with enhancement. It is often astounding. Dark lanes and regions within the globulars often become substantially more prominent, posing the age-old question of their makeup even more assertively. (Are they present within the cluster itself, or merely interstellar dark obstructions somewhere in the line of sight between us and the cluster?) Regardless, they are a never-ending source of interest and speculation when viewing those clusters in which they seem to be present. They usually demand the most attention as you view the main subjects themselves; subjects for discussion will always be what can be seen, how well, and at what power they show best.

Those examples of globular cluster presently caught up in the gravitational pull of the Milky Way (but not yet part of it) are also highly interesting viewing challenges. Some other clusters actually belong to nearby external satellite dwarf galaxies, which are themselves caught in the Milky Way's tug-of-war. Either way, the globulars in question are all quite remote with respect to us on earth. It seems always that the quest is to resolve as many stellar points as possible in these relatively distant objects; image intensifiers in particular, with their fine resolution, may make that pursuit more effective. Globular clusters belonging to major independent galaxies are a much greater challenge still. They are seen as barely more than stellar points in amateur scopes, if indeed they are visible at all. Many observers enjoy the challenge of being able to detect these external clusters in some of the closer galaxies, if you regard millions of light years as close. To my thinking, such activity, while understandably fascinating and compelling, does not equate with the objective here of finding great real-time spectacles of the night sky. However, I would not want to discourage the search, and certainly an image intensifier will make it much more profitable. The Great Galaxy in Andromeda is literally studded with them, having almost three times as many as our own galaxy. You will need to consult detailed star charts, with cluster locations clearly marked, in order to take part in the hunt. Be sure of their accuracy, a well-known bugaboo!

Results in the Field

Under dark skies, image enhancement provides a benefit with these subjects in a somewhat different manner than you might expect. Actually, many star clusters reveal themselves in clear skies from suburban locations pretty well. Although darker surroundings make possible the resolution of still fainter stars, the gain in the number of stars is not likely to be nearly so significant in real-time as one would suppose, compared to those that were seen at equal magnifications from suburban locations. This is because the bulk of the population of many of these clusters frequently consists of an abundance of similar stars, comparable in size, magnitude, and type. In the clusters well seen from suburbia, these very stars may be well within the magnitude limits imposed by poorer locations. The all-important keys, however, are the increase in brightness of those stars that we do see, as well as the far greater manifestation of the true span of the cluster as fainter peripheral stars become visible. The magnifications that sometimes become possible may allow us to crack these clusters wide open, often to the core! I can use my 5X TeleVue Powermate Barlow on the brighter subjects, and it is with this that the advantage of enhancement becomes readily apparent, as the subject is able to retain a good deal of brightness.

Globular Clusters

You already realize that it is because of their stellar densities and relatively small apparent size that globular clusters will likely grant our most spectacular results with enhancement. One of the best examples for a truly dramatic comparison to the views we may have had from suburban confines would be that old perennial, M4 in Scorpius (Figure 5.24). Not prominent enough to make a monumental statement from my own suburban location, this particular cluster was relegated to the supplementary second catalog in my previous writing, *Visual Astronomy in the Suburbs.* In the wilderness it provided one of the biggest surprises among all of the globular clusters.

Viewing M4 under dark skies, it is hard to say anything but, *wow!* Now the gaps that made it look so sparse from the suburbs are all filled in. However, because M4 is relatively open, we can still see a good degree of stellar separation, and the great cluster does not merge into one concentrated glowing ball as do so many others. Along with this, enhanced viewing produces a greatly exaggerated view of the chains and loops for which the cluster is so well known. This makes the view quite extraordinary and seemingly unique among globulars. Newly prominent is the great central "bar" (a bright and straight chain of stars bisecting the hub); the brilliance of the whole, in contrast to all the tiny dark separations between the stars, is all the more striking. The view at lowest power is quite sufficient to reveal all of these attributes, while retaining the greatest sharpness and image contrast.

In fact, the view, even with low power through my image intensifier, is so magnificent and resolved that M4 went from being a so-so object to one of the

Figure 5.24. M4.

best and most imposing. Why is this? Perhaps it can be best explained by the comparison with so many other globulars. First of all, it is really close to us, astronomically speaking, at only(!) 6,500 light years. The majority of globulars are much more distant, denser in star populations, and therefore more compact in their apparent dimensions; these can perform well through the telescope under less than optimum conditions. M4 is just too spread out to do as well in the same situation, since many of its smaller stars are just too dim to fill in the gaps satisfyingly.

Always wonderful to see, of course, is legendary M13 in Hercules (Figure 5.25a). This grand sight is the largest of all the Northern Hemisphere globulars. Appearing significantly brighter and fuller now than the view from the city, it nevertheless doesn't show such a striking change as does M4. However, the famous "propeller lanes" which I wrote so extensively about in my previous book are now so easy to see that it is no longer necessary to provide a chart to be sure of their location! In fact, these lanes also are apparent even in the conventional view if one knows where to look, so I cannot understand why they were ever considered absent, needing "rediscovery" in the 1970s. The reason they seem to hide on so many photographs from earlier times is not, in fact, because they are completely dark and all the light from stars behind them is blocked. You will notice what actually happens as the length of photographic exposures increases: the light from more stars within the lanes themselves registers and fills in the void. This is the main reason for their disappearance in photographs, although the total light from the cluster also begins to spill into those regions on long exposures. However, our enhancing devices only seem to exaggerate the appearance of the lanes in live viewing, at least with apertures up to 18 inch, such as mine.

Again we have some valuable comparisons with the next two examples, both Astrovid StellaCam EX video images taken through two different apertures. (See Figures 5.25b, at 15 cm, and 5.25c, at 25 cm.)

Interestingly enough, these images provide very revealing indications of some of the relative strengths and weaknesses of both imaging methods, that of single intensified frames versus integrating video cameras. Personally, I find intensified video images to be the more revealing and subtle, although they are a little less

Figure 5.25a. Great Hercules Cluster M13.

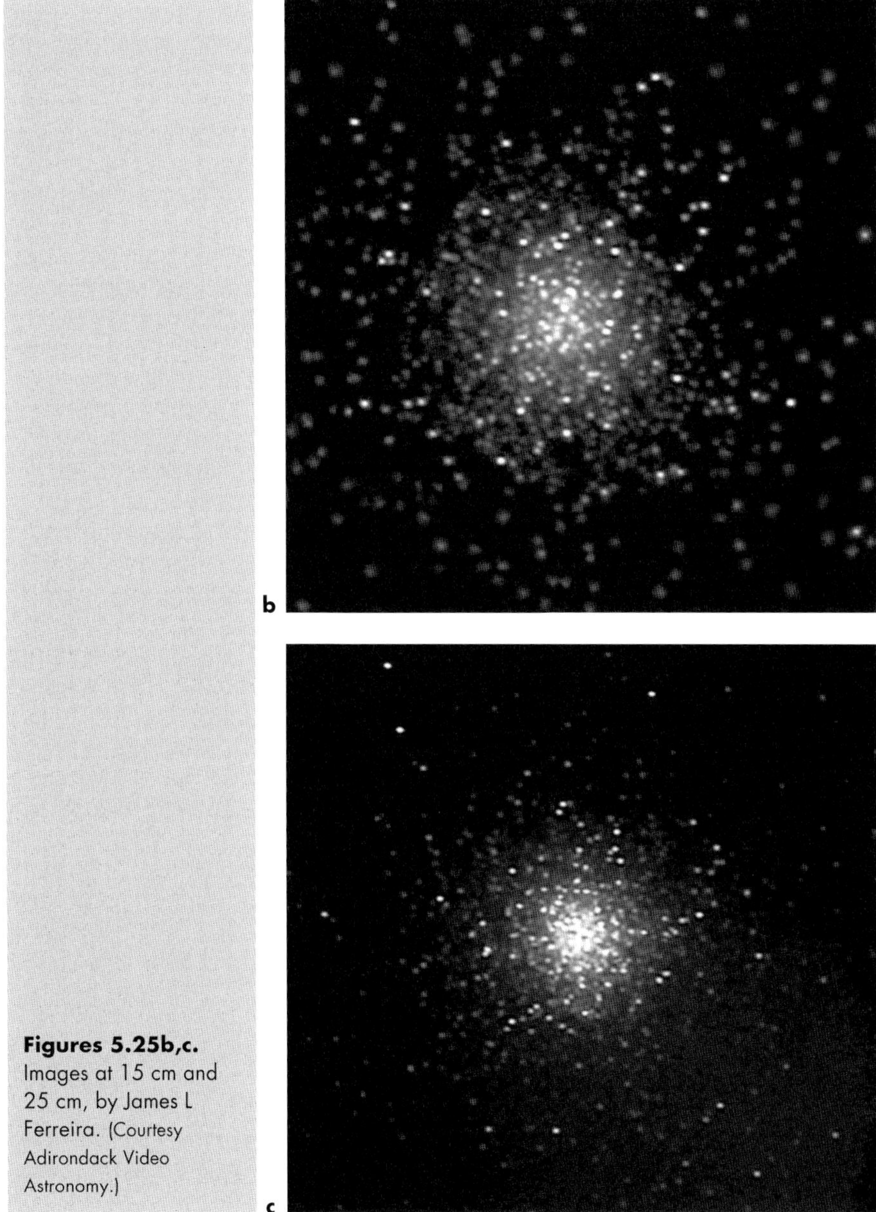

Figures 5.25b,c. Images at 15 cm and 25 cm, by James L Ferreira. (Courtesy Adirondack Video Astronomy.)

defined. The "propellers" and other dark features in other clusters stand out much more starkly in the intensified frames.

The near-perfect symmetry of the outline of M13 only underscores its position as one of deep space's premier objects for live observing. However, intensified viewing brings out many irregularities in illumination, which draws attention to its varying stellar makeup. There now appears to be a bright stellar "X" at the center of the cluster, and the often-described "spider-like" appearance of this cluster most aptly sums up its outline. "Spiral chains" of stars are also commonly alluded to by many observers, but these aspects actually may be more a feature of the mind imposing its own patterns on otherwise random groups of stellar points than true characteristics; you decide. At least moderate power will be necessary for the best views of M13, since it occupies significantly less space than its more spread-out rival M4, and is comprised of much closer stars; it has easily sufficient illumination to withstand even high magnifications.

Although ranked as one of the Milky Way's great sights, M22 is sometimes a subject of some disappointment for many Northern Hemisphere suburban observers. Regrettably, it lies at a fairly unfavorable position in the north; it is, after all, a southern cluster, and as such, may be placed too low in the sky for optimal access. Nevertheless, for those fortunate enough to live at latitudes near enough to the equator to see it well, and of course, for those observers who reside in the Southern Hemisphere, M 22 will prove to be one of the most spectacular clusters in the sky (Figure 5.26). Indeed, from my own home location in southern California, it is somewhat more imposing even than the great cluster in Hercules, M13.

Actually, in terms of its true dimensions and stellar population M22 is not exceptional among globulars, and could be described as quite average, particularly when compared to its other grander brethren. It appears so magnificent and large only because of its close proximity to us, at something less than 10,000 light years, a fact that not only presents it an object of considerable magnitude and size, but also, like M4, gives us the advantage of ready stellar separation. In any form of viewing it draws attention to itself at the first glance, and even in the smallest of apertures that the amateur is likely to use, its stellar members will begin to resolve. Enhanced viewing turns it into a vast, somewhat oblate, luminous ball of unevenly dispersed stellar points. The stars within M22 look like random placements to my eyes, certainly with my image intensifier; though often described as appearing in chains and streamers within the cluster, the stars comprising this awesome sight appear in a studded blotchiness, certainly not typical of more distant globulars. However, striking straight lines of stars "shoot" outward all around its circumference, a feature more noticeable here than in most other globulars. As with most examples of this type of cluster, there also appears to be a degree of dark obscuring matter scattered around the core region

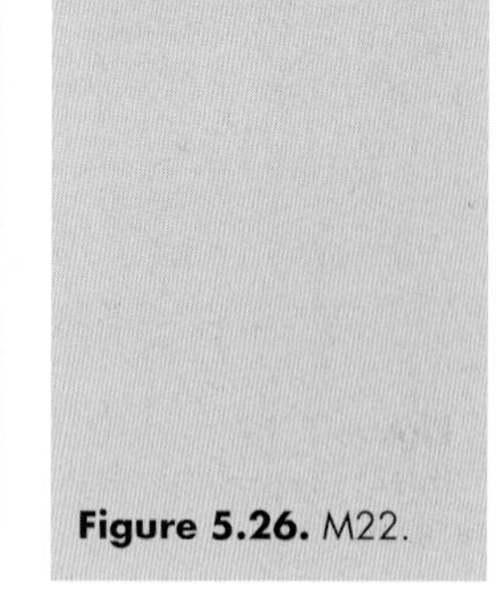

Figure 5.26. M22.

of M22, made all the more visible in enhanced viewing devices with infrared sensitivity. Because of its magnificent and well-resolved appearance in the telescope, it rates among the most interesting globulars, justifying frequent visits.

The cluster M5 in Serpens provides another interesting comparison with what we were able to see from the suburbs (Figure 5.27). It is a magnificent specimen

a

b

Figure 5.27 a. M5. **b.** M5, as seen from the suburbs.

under any type of viewing, ranking alongside only a few other top luminaries in the Northern Hemisphere. From my home base, with my Collins intensifier, it is routine to make out a dark rift with tributaries, slicing the cluster into two unequal sections. This feature is as striking as it is unusual, and not commonly described. On one side of the lane, the larger section easily outshines the smaller, not just by nature of its size, but also because of the brighter stars on this side. In dark skies, for much the same reason that photographic exposures often fill in the "propeller lanes" of M13, this rift is now largely blotted out by increased resolutions of stars within the corridor of darkness we used to see. It is still present, of course, and is visible now as gray division, but it is now not nearly so significant a feature for observing.

Instead, we now become more aware of another dark lane crossing the great cluster at right angles to the former, and curving around and within some of its outline. It is easily visible because from our dark sky site we are able to view a greater width of the cluster, made up of fainter stars against the surrounding space. Additionally, the intensity of the core is stunning; it does not gradually get brighter as we work our way inward. A wide region of much reduced stellar density around the bright central blaze itself seems to suggest something of a spiral formation more than do many others. If you take the time to examine the two examples below from this interesting comparison of viewing locations, the differences will be very apparent.

Another old favorite, M3 in Canes Venatici, exquisite in any form of viewing, shows new brilliance under true darkness, and is one of the most impressive globulars of them all (Figure 5.28). More concentrated yet than M5, it has a much more conventional appearance, being both more tapered in brightness and with fewer stragglers on the outskirts. It is somewhat oblate in shape, and has a fairly even central brilliance, although small dark lanes may be seen crossing the region in arcs. I have seen many time exposures that are no more brilliant than the real-time view here, and in fact, reveal substantially less detail in the core region. The spectral sensitivity of the Generation III image intensifier is certainly partly responsible for providing us with new insights in globulars, as it emphasizes

Figure 5.28. M3.

Figure 5.29. M10.

more the stellar makeup of their individual spectral characteristics, which typically radiate red and yellow wavelengths. I don't believe other enhancing devices are nearly so sensitive to these aspects, and as such, are less likely to show M3 so wonderfully.

There is hardly a limit to the amount of exploration that can be done with the Milky Way's globulars. Too numerous to feature more than a few here, you will find that the variety of form and appearance is quite broad; they are all individuals. Some of this is due to the structure and age of the cluster itself. Frequently, big differences in appearance are due to the effect of the cluster's distance; this produces a wide range in the brightness and separation of the stellar points. M10 in Ophiucus is an example of a globular that is quite close astronomically speaking (at around 15,000 light years), but also not especially large (Figure 5.29). Despite, or perhaps because of, the lack of really bright stars within its midst, it allows for much separation and resolution. Its stars are relatively uniform in prominence, and a telescope of modest size, excluding truly small apertures, will allow successful resolution across its structure. To a lesser degree this can be said even when using conventional viewing, where the cluster has a remarkably unusual appearance of tiny stellar points, to my eyes like sparkling grains of sand. The enhanced approach produces a view much more dazzling and resolved still, though M10 nevertheless is not so striking as that of grander globulars.

A most interesting globular, M54, is a prime example of one external to the Milky Way (Figure 5.30). In fact, it belongs to the Sagittarius Dwarf Galaxy, a small satellite of our own system, and is apparently the actual core of the galaxy itself. Comparable in size to the great and colossal Omega Centauri, its likely role lends additional weight to the often stated theory that Omega Centauri is, in fact, the surviving core of another galaxy. Because of its great distance, at almost 70,000 light years, M54 is usually seen as just a bright and tiny smudge, unresolved in amateur telescopes. Descriptions often mention only a "granular" texture or outline, even through the larger apertures available to amateurs. However, under image intensification and just a moderate aperture, M54 begins to reveal its stellar structure, and on a steady night begins to give us one of the

Figure 5.30. M54.

most amazing revelations such equipment can bring to star clusters. With larger apertures and high magnifications, increasing resolution is possible, although the bright core region remains a glowing ball, owing to the staggering density of stars as seen from such great distance. However, I have seen a surprising number of stars resolved within the distant outline of the cluster.

You will need nights with the steadiest air for this one. As of yet, I have not been fortunate enough to capture the finest resolutions I have observed on moving videotape. In this particular instance of deep space viewing, actually seeing something of the stellar resolution is not at all unlike trying to make out subtle planetary details; higher powers may be found quite necessary. Look first for the granular appearance all around the core; gradually, individual stars will begin to resolve as you work your way ever inward. I find the splitting of at least some of M54 to be an engrossing pursuit. One can only imagine what a magnificent sight it must be at closer range.

Open Clusters

Most of these gems of the sky probably should be dealt with in an entirely different book and context. In fact we have already seen some relatively compact examples of open clusters contained within the great nebulae, and providing the light or energy by which these gas clouds shine. The problem the grander and more famous examples pose for telescope users is the great extent of the sky that most of them cover. They are usually too wide and spread out in stellar populations to be effectively seen other than in the wide fields of smaller apertures and binoculars, or better yet, by "richest field" telescopes. Even the nonmagnifying lens component for our enhancing devices will tend to render them visually quite uninspiring, as part of their charm is the variety of colors and brightness that different clusters show. As a specialized field, there is no limit to what is in store for the ingenious amateur, willing to invest the time to explore to maximum effect the realm of open clusters that have the most potential. When selecting eyepieces

Figure 5.31. The Wild Duck Cluster M11.

for very low powers, be sure to consult reliable manuals concerning optical design; too low a magnification will not take advantage of the full aperture available and may even introduce detrimental effects. This, in itself, restricts one's options when it comes to selecting a suitable telescope.

The Wild Duck Cluster M11 (Figure 5.31) is so famous among amateur observers that one would be forgiven for assuming it is a globular cluster. In fact it is one of the finest and densest open clusters available to the astronomical telescope user, and so compact in the view it presents that it does indeed rival many of the well-known globulars.

Impressive, even from the suburbs, it further comes into its own under dark skies as a considerably more brilliant object. Structurally, it is particularly interesting, and the features so aptly described by Admiral Smythe, who penned its name long ago, are readily visible in the telescope. Look for the front line of "birds" to the east, a V-shaped formation that seems to lead the flock. Though the cluster is set against a backdrop of faint stars, this feature, often difficult to see with conventional viewing, is an easy mark with enhancing devices. The considerable unevenness in distribution of the stars is also quite striking, something possibly caused by dark obscuring matter. You will find much of interest to study in M11.

So now we take off in our celestial spaceship of the imagination until we reach the edge of our neighborhood in space (all that comprises our own Milky Way Galaxy), beyond the very last star we can see. Looking ahead are countless billions of other galaxies across the great void separating our system and them. We will have to travel inconceivable distances to reach even our nearest neighboring systems of the local group. The sobering thought is that despite the incomprehensible immensity of our own galaxy, we must confront the fact that it occupies just one tiny corner of the cosmos. So it is toward the other galactic star systems that we now turn our gaze and wonder.

CHAPTER SIX

The Great Beyond

Galaxies

Less than a century ago, all the stars belonging to the Milky Way system were considered the universe. It was believed that nothing lay beyond its confines, and it existed as a complete entity in space, our "island universe." The term "galaxy" was used to describe the entirety of creation.

It was too frightening to think for a moment that most of the so-called "nebulae" were actually other galaxies, and that the universe was an infinitely grander and more overwhelming place than anyone had ever dared to imagine. As the true nature of galaxies began to be known to cosmologists such as Edwin Hubble, it struck terror in the hearts of many, and rightly so. The universe is staggeringly vast, beyond human capacity to comprehend. The lack of being able to adequately process its scale in our tiny minds gives us pause to reflect on our true place within it. Humans have had to see themselves in less preposterous terms.

Even 50 years later, when the discovery and confirmation of the true nature of these galaxies, "island universes," had become old news, many astronomers still found it difficult to refer to them as such. We can still read in books from the mid-twentieth century the terms "nebula" and "spiral nebula"(!) being applied to them. This was long after the debate was over. I still remember the closest full spiral system to us, the Great Andromeda Galaxy M31, being referred to routinely as the "Great Nebula in Andromeda," as recently as the 1960s. Believe it or not, I saw it referenced in that manner in a recently published article. It does seem, nevertheless, that virtually everyone has finally learned to live with the reality of the cosmos. It also seems rather quaint that we find ourselves referring to any of these other immense island universes as "objects," for want of a better term!

Galaxies are prime objects to study (there is that term "objects" again), in most ways, to a greater degree than any other in deep space. When we are engaged in viewing them live this is even more the case, especially if we turn to our new secret weapons, image intensifiers or advanced video devices. Galaxies offer a great variety of sights, and in live viewing you never know when one of them will take your breath away. With the equipment widely available to amateurs today, it is humbling to think that, in the nineteenth century, it took all of Lord Rosse's great 72 inch (1.83 m) at Parsonstown to reveal the spiral makeup of galaxies for the first time. How fortunate we are today; it is now routine to detect such structure in apertures barely big enough to qualify as that telescope's finder! Some of this is due to superior optics, some due to awareness of what is there, and a further component being the high-quality and sophisticated equipment and accessories that even amateurs now have at their disposal. Of course, none of what we see can be reconciled in any way with the knowledge of what these "objects" actually are; it still makes my head hurt! First we must accept the reality of what we see, and the almost immeasurable distances to them. The space that separates us from them can only be measured in absurdly huge units of time, time becoming the all-important gauge of distance. In relativity, time relates directly to space and its three other dimensions. For our purposes time and space might as well be one and the same, since it is time that ultimately separates us from other entities, maybe across vast voids in the cosmos.

The light from all of the universe's galaxies has traveled for millions or even billions of years to reach planet Earth. On a tiny scale in the cosmos, our own sun, just an average to below-average star, is the better part of a million miles in diameter. The next nearest star to us, Proxima Centauri, is over 4 light years distant, and one of "mere" billions of stars constituting the Milky Way Galaxy. When we are looking at any one galaxy across space, the faint glow that meets our eyes is all that remains of the light of its own billions of suns. It is difficult to maintain an awareness that the dimensions of such a "little" galaxy could exceed even 100,000 light years; in the field of view, it appears so modest, benign, almost gentle. Surprisingly, even the entire visible stellar makeup of our closest spiral galactic neighbor, M31, appears as no more than just a bright fog in the eyepiece. The "fog" is the fire of its billions of individual suns, each separated by staggering distances in themselves; we see them only as a completely unresolved, blended cloud. Surprisingly, enhanced viewing does not reveal the galaxy's makeup in the way we might expect and hope; while the well-known dark lanes show well (as they also do with conventional viewing), the main benefit is only the much more dazzling blaze of light that now lights up the field of view.

It does not take a lot of thought to realize that if that is all there is to see, even given its neighborly distance (a paltry(!) 2.4 million light years), we should therefore have cause to contemplate the unimaginable scale of what lies before our eyes. It requires large telescopes and time exposures to break down M31's structural makeup into individual stars. Coincidentally, it was this particular galaxy that was the subject of Hubble's first such investigations at Mount Wilson Observatory almost 100 years ago, his photographic plates finally showing the true stellar makeup of the extragalactic "nebulae." Yet our galactic neighbor is merely the closest of billions (yes, that other term again) of other galaxies. The term "billion" seems to be a standard numeric description in the universe, yet it is pathetically inadequate to put things in a perspective that is in any way mentally palatable. Just knowing what constitutes these galaxies, and the possibility of countless life forms within their own confines, dazzles the mind. If other civilizations exist within these other island universes, then these are where their own perceptions of the cosmos exist, as they look outward through their own galaxies' star fields and constellations. To them, our immediately visible part of the universe, the Milky Way Galaxy, is not perceived as anything other than just another telescopic smudge against the background sky. To us, the Milky Way Galaxy is a dazzling showpiece of thousands of spectacles within its rich and diverse makeup, and might as well be all that there is.

To extract the image of the central region with two lanes faintly visible here in Figure 6.1a, it was necessary to suppress the brightness, which otherwise tends to overwhelm anything else we see in enhanced lower power views. The six-second exposure at right (Figure 6.1b) gives a better representation of the live intensifier view of the lanes, as well as this region.

Being so bright, M31 is eminently capable of withstanding increased magnification, and further detail within the dust lanes may be seen. The most

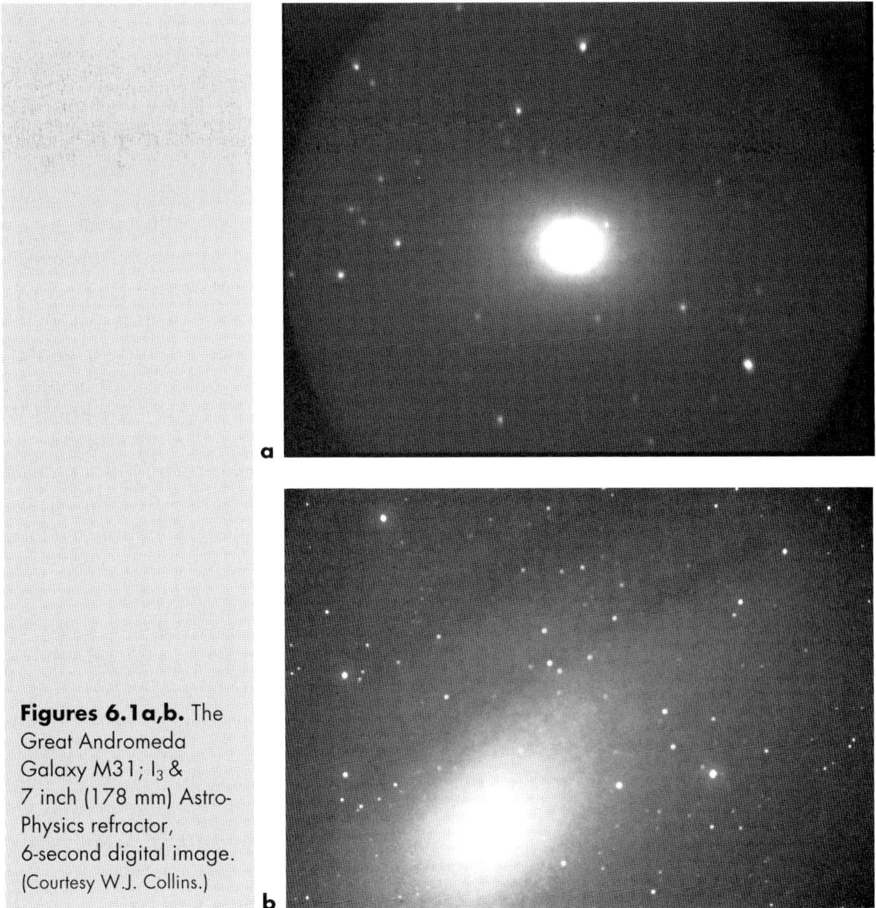

Figures 6.1a,b. The Great Andromeda Galaxy M31; I_3 & 7 inch (178 mm) Astro-Physics refractor, 6-second digital image. (Courtesy W.J. Collins.)

Figures 6.1c. Dust lanes in M31, as imaged by James L. Ferreira, StellaCam II. (Courtesy Adirondack Video Astronomy.)

advanced integrating video cameras are particularly suited to imaging them effectively. (See Figure 6.1c.)

Seeing galaxies from dark sky sites in conventional telescopic viewing, without any form of enhancement, can be a sufficient thrill for most observers. With image enhancement, the visual thrill will often turn to stunned amazement. Many familiar galaxies become more impressively revealed, and countless others now come within our grasp. These galaxies new to our view will probably appear now as did many of the brighter galaxies in conventional viewing. Nevertheless, under dark skies, usually it is galaxies that are already striking in conventional viewing that offer the most potential for dramatic revelations in real-time enhanced views. These galaxies are likely to belong to our own local galaxy cluster, or to another nearby group. Those contained within some of the other great galactic clusters, such as that in the constellation Fornax, may not necessarily offer greatly increased visible detail. This is because of their much greater distance from us, and unless simply seeing them more brightly is a sufficient goal, they may even be disappointing. Therefore, successful viewing depends on several things: what your viewing objectives are, specific characteristics of any particular subject, and the exact distance at which it lies. (Incidentally, the galaxies making up the Fornax Cluster stunned me when I first viewed them with image intensification, even from the suburbs; it seemed galaxies were bursting out of the sky everywhere. The density of the spectacle was the distinguishing feature, of course, not the individual resolution of galactic form.)

With all forms of astronomical observing, learning how to extract the maximum present in the image is of key importance. Generally speaking, we cannot pretend in any form of live galactic viewing that what we will see represents a complete face of these structures. However, it can give us more than an inkling. In some instances, particularly with the brighter and better-resolved galaxies, the intensified or otherwise enhanced view will be so close to the well-known observatory portraits that there is little else to add. However, the real-time views we will have are still unlikely to be quite as brilliant and contrasted as these grand telescopic images, so we must learn to allow our eyes and mind to make these adjustments for themselves. This is where the eye comes into its own in live viewing, and is all the more apparent on an image intensifier's phosphor screen than the view on any monitor, regardless of the type of video camera we employ. Even so, most typically the most spectacular subjects will still require that we stretch our eyesight to the maximum in order to see all that is there. The key here is that enhanced viewing will likely make our job much easier, and when it is applied to the best-suited galaxies, the results are quite remarkable. It may also bring out regions that otherwise would hide, due to the specific wavelengths of light the particular enhancing device we are using is emphasizing. However, less-suited examples will be revealed in ways that add nothing except brightness in the central cores, reducing the eye's awareness of other regions; you may even prefer the conventional view.

The majority of the galaxies featured in this book are well within the grasp of most amateur scopes, and they also provide some of the best potential for spectacular enhanced viewing. Whether they will jump out like an observatory portrait is always the question in searching for the best subjects. You already know that usually this distinction will be reserved for the relatively nearer and brighter subjects, our neighboring systems. However, by this we can include a large catalog of galaxies lying within a 50 or 60 million light-year radius. Their potential for live enhanced viewing also depends on the specific characteristics of their structure and angular placement relative to us, as well as the specific wavelengths of light most favored by an intensifier or other enhancing apparatus. Let us review these important viewing factors and how they relate to galaxies themselves.

Edge-on spirals are almost always prime targets for enhanced viewing, and usually the least likely to disappoint. Seen from the side, the dusty regions of their spiral arm encirclements (which, to our view, are concentrated and compounded) as well as the many old population stars of their central hubs provide rich infrared and red spectrums. These are wavelengths that image intensifiers, in particular, thrive on, of course. The StellaCam EX is also quite responsive in this spectral range (though I imagine to a lesser degree), and should also favor edge-on systems to some extent. Resolving detail on these subjects is always an exciting prospect, and many such galaxies oblige us fairly generously. It is often dark encircling lanes, often present in these galaxies and now seen from a side-on perspective, that offer a huge bonus; they are not nearly so prominent in conventional viewing.

Textbook examples of bright, face-on and near face-on spirals can be frustrating just as often. Their spiral form is now facing toward us from above or below, often presenting a fully open view of the galactic arms, but they may be disappointing when viewed through the telescope. In conventional viewing, these objects may be too spread out, or simply too remote and faint, to easily discern their structure; the unskilled observer will probably find many such galaxies profoundly unrewarding with such viewing. However, it is these galaxies that also provide some of our most unexpected potential with image enhancement, but this is ultimately dependent on the spectral "fingerprint" of each galaxy. As you know, intensified, and indeed any form of enhanced viewing has special and unique characteristics depending on the particular response of these devices. Intensified viewing, in particular, presents especially different responses to the eye's normal range and emphasis.

To go further, in any given galaxy, the young blue stars making up the bulk of the spiral arms are usually quite faint relative to the central core, which consists of older yellow and red stars. Theoretically, at least, when it comes to enhanced viewing, and maybe more so with intensified viewing, the blue stars' spectrum will not be favored by these devices' response. This means that the more favorable spectrums of the central core will usually be enhanced relative to the spiral arms. We would suspect therefore that face-on spirals will always show these enhanced central regions prominently, leaving the bluer, relatively nonenhanced regions even less apparent. Our suspicions are often well founded. In any case, it is also true that sometimes their face-on aspect makes them just too spread out for such orientations to provide a bright enough form to discern spiral detail anyway. This is even more noticeable where blue light is predominant in the arms. However, do not become too easily disillusioned by my tempering of your expectations; most of my greatest surprises and revelations have been with face-on galaxies! This is apparently because strong red and infrared components within the spiral arms themselves can indeed render them quite visible with enhancing devices under truly dark skies.

With face-on galaxies we will need to experiment with the three primary galaxy types; some offer us greater prospects than others. These are characterized by the designations Sa, Sb, and Sc. There also exist various blends of one type with another, such as Sab or Sbc. The further designation A (i.e., SA) refers to the absence(!) of barred structure, and is often used in galactic listings; the designation B for the galaxy type (SB) indicates some barred structure. Do not confuse the uppercase of these two letters with the lower, since they mean entirely different things. In suburban locations, being aware of galaxy type is virtually moot, since spiral structure is likely to be obliterated by sky pollution of various forms, no matter what. But now, in true, clear, darkness, we will unlock a new capability and potential.

The central cores of all of these galaxy types will always be enhanced with the types of image enhancing equipment we are discussing. This core enhancement ranges from greatest in Sa (early type) galaxies, to least in Sd's (late type). Overall, an Sa galaxy with a prominent central core, and characteristically more tenuous arms, is likely to show only a brighter core – seemingly at the expense of the arms. These galaxies usually appear as having just a brighter hub overall, along with correspondingly much-reduced "halo" detail. Some Sb types, quite plentiful in the universe, however, present us with some better opportunities for enhancing devices. Classic Sc's and Sd's tend to do even better still, though they often show low total surface brightness in their relatively evenly spread out form. However, these galaxies' well-developed dense spiral arms, often full of emission nebulae from young, hot stars, are likely therefore to radiate prime HII wavelengths from the intensifier's standpoint, and can produce fine spiral views, especially with moderate to large apertures. I have found that this is where the intensifier provides some of its biggest surprises in dark skies. As always seems to be the case with enhanced viewing, the overall galactic halo is reduced in the view, but the bonus is that the spiral structure may well be significantly enhanced instead. Again, experimentation is the key.

Barred face-on spirals (SB galaxies) follow the same guidelines, though the presence of the central bar gives us new potential for visible detail, which is often well displayed. The central bars contain considerable quantities of stars radiating the frequencies that do best in enhanced viewing. However, quite common among galaxies is the designation SAB, which refers to a galaxy with some traceable bar structure. Add to that a, b, c, or d in lowercase letters, alone or in combination, and you have the same normal range of galaxy designations as before. Many galaxy "bars" become quite significant in enhanced viewing, along with any of the other possible features.

Elliptical galaxies are designated as E0 through E7, E0 being virtually round, with E7 being the most elongated. Similar to them are lenticular galaxies, S0 types, which are sometimes hard to tell apart from ellipticals, but differ in that they show some core structure, and sometimes even encircling dust lanes. Lenticulars also sometimes show something of a barred structure, and hence are designated SB0 types. There are other letters and abbreviations sometimes used as well, depending on who is writing the classification, but what we have here is sufficient for all intents and purposes. Elliptical and lenticular galaxies, consisting of red and yellow stars, almost always are greatly enhanced, although these subjects are not usually among the most intriguing visually, or even likely to show detail. Indeed, if they appear to do so, they are actually more likely to be an "early" form of spiral instead. In any event, of the two types, only edge-on lenticular galaxies will reveal dust lanes. However, as always, it is the exception we are looking for. Such is NGC 5128 (Centaurus A), apparently a large elliptical combined with a swallowed-up spiral. It proves nothing less than breathtaking in the enhanced view; speculation persisted for decades about its true nature. The giant elliptical M87 is famous for its jet of matter, shooting directly outward. Seeing this jet is a prize for live viewing and larger apertures, made all the more possible by use of new approaches, although I cannot lay claim to having seen it yet. Those who have seen it, and with conventional viewing at that, comment on the high magnifications they have employed, as well as the clarity of the skies and the deep attainment of dark adaptation necessary. Since the galaxy is not among the brightest ellipticals, it may require more viewing skill and patience than I have been able to muster up to this time. One day perhaps I, too, will lay claim to a sighting, hopefully made easier by image enhancement.

Countless galaxies of all types are presented to us neither face-on nor fully edge-on: so they appear to us partially edge-on. This is a natural and most expected state of affairs, considering the vast galactic distribution throughout the cosmos. Many of these partially edge-on galaxies are among the best for electronically enhanced viewing. Certain irregular and "peculiar" galaxies are also capable of providing us with some spectacular sights. Regardless, our enhanced viewing experiments provide a fine alternative to conventional viewing, and do present a brighter and different manifestation of galaxies. It seems astounding to me to be able to show fine features of obliging galaxies to other people, and with such ease, particularly to those with little or no viewing experience or skills. Relative to conventional viewing, the detail can be both striking and easy to see.

With just a little careful analysis of what we are about to look at, it is possible to have at least some idea of the manner in which any galaxy will respond to either form of viewing. So here we are, relocated to the big, dark sky. It was here that I became fully aware of the vast potential of my image intensifier, as well as image enhancing devices in general. But always be prepared for the unexpected, especially when you do have the time to observe everything possible in any given part of the sky! Below are some of the results I obtained with my favorite deep space subject: galaxies. It is my hope that these may also point you to some of the great potential that I have found.

Results in the Field

The greatest single difference between suburban viewing and that at dark sky sites is the great ease with which so many deep space objects may be accessed, even with conventional viewing. The downside of utilizing today's advanced enhanced viewing systems on these same objects is that while we will often gain greatly increased brightness and contrast, the new electronic manifestation often reduces the completeness and subtlety of our subjects. The somewhat inexplicable sense of three-dimensional structures hanging in space, the general glow of the "halo" forming the outer regions of galaxies, along with all kinds of subtle and highly delicate details, sometimes do not lend themselves to an enhanced electronic image. However, other aspects that may have remained invisible in conventional viewing, or less than ideally seen, may instead jump out in stark relief. Other negative aspects do not necessarily translate to all electronic views. The brightness of the object being viewed may also give it an impact that can scarcely be expected in real time; contrast, resolution, and increased structural attributes also are frequently much more apparent in the intensified or otherwise enhanced view. It is here, though, that the special value of live viewing with image intensifiers stands on its own, in my opinion. Such direct and effective viewing is able to bring a closer realization of subtleties than any view on a monitor or by camera.

Galaxies Revealing Spiral Structure

Seeing spiral structure in spectacular fashion, as well as other structural galactic features, probably remains my biggest reward among all deep space viewing! For the most part, in the suburbs, such an exciting prospect tends to elude us in real time, no matter how we approach it. It is one of the greatest paybacks for hauling our gear to remote places to finally have the opportunity to witness such things with our own eyes. We know that the use of enhancing devices only increases our viewing prospects. Good results with many less than likely subjects are all the more impressive to me, especially since I had been prepared to anticipate some degree of disappointment in this particular quest. Many galactic features are contained within areas of blue light emission, and so actually seeing them clearly showing in real time was not something I was counting on. That such results are often stunningly possible with today's enhancing systems at dark sky sites, on apparently unlikely subjects, is good reason to remain more optimistic than the official word would lead us to anticipate!

Certainly, tempered expectation is typical of the mainstream buzz concerning the application of image intensifiers. Even Bill Collins does not seem to claim in his descriptions of the performance of the I_3 (or modern image intensification devices in general) the level of grand assessments in viewing galaxies that I describe. However, my descriptions are based strictly on actual results I have experienced with my own eyes, so I can only ascertain Collins does not wish to oversell his products' capabilities. More power to him; there are going to be a lot of unexpectedly happy people. (He seemed delighted with my reports and images.) But first, people must try it!

It is when we realize the unique attributes of each approach of viewing, conventional and enhanced, that we may begin to enjoy the best of both worlds. For me, one of the greatest thrills with intensified viewing in dark skies was the revelation provided by face-on Sbc-type M51, the celebrated Whirlpool Galaxy in Canes Venatici. Here is the grandaddy of them all, the first galaxy to reveal spiral galactic form. This revelation was granted to Lord Rosse in the mid-1800s, using his 72 inch reflector, then the largest telescope ever constructed. In my previous book, *Visual Astronomy in the Suburbs*, I commented upon my regret at not being able to include this celebrated galaxy among prime objects to view in suburban skies; it just did not provide a spectacular view, the structure being hard to detect. I was sometimes able to make out a trace of its spiral structure in conventional viewing on transparent nights using a narrowband filter, but spectacularly so? Unfortunately not. With intensified viewing, galactic form disappeared completely. What a change was in store for me at dark sky sites! (See Figures 6.2a and b.)

Conventional viewing from dark locations shows M51 as a bluish structure immersed in a glowing halo. Seeing the familiar spiral form spinning outward from the central hub of this legendary sight is now quite routine, even in moderate apertures. However, nothing prepared me for the intensified views with which I was greeted out in the desert. I was ready for disappointment, especially since the galaxy is a rather spread-out example of a face-on type, and the arms are well documented as very blue in spectral characteristics. One clue to success, though, was in its being an Sbc type, a better match than many galaxies to our enhancing devices' response. The other clue is the presence of HII regions in the arms themselves; these are strongly apparent in pictures taken from space. I knew I had at least a fighting chance of seeing something! In conventional viewing, the general glow of the galactic halo actually merges with the spiral form to some degree. This blurs their separation visually. In the enhanced view, the "halo" part of the equation is expectedly absent, and instead, the arms take on a new and stark appearance. Now these famous spiral arms will jump out, almost in relief, at first glance, the spiral aspect becoming foremost in the view. Even the bridge that seemingly connects the large galaxy to its smaller neighbor, NGC 5195, is easy to see; the slight bar shape of this smaller companion galaxy

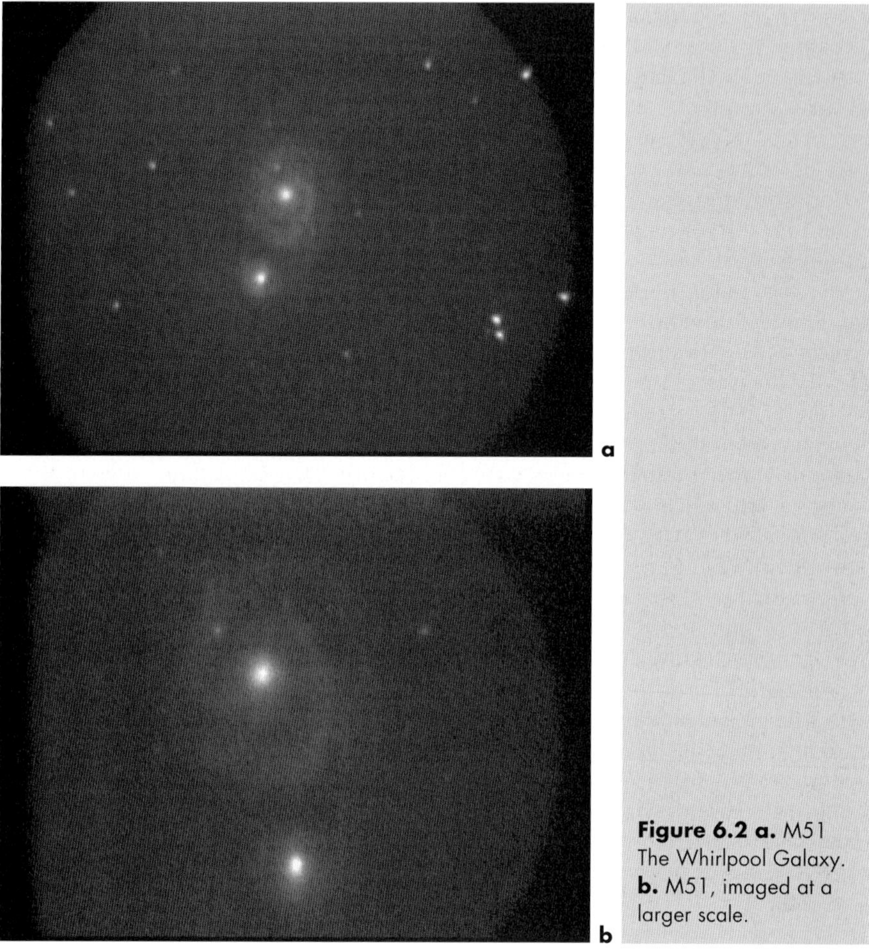

Figure 6.2 a. M51 The Whirlpool Galaxy. **b.** M51, imaged at a larger scale.

can also be detected, as well as the slight dark obscuring lane in one of the arms of M51 crossing in front of it.

Compare what you see with a good photograph of the galaxy, and you will notice many parallels. Mottling in the arms, and streaks "peeling away" from them, are also apparent. Except in the greatly reduced brightness compared to what one would expect to see in a photograph, M51 looks amazingly complete in detail and form. Even on a video monitor, the view is still quite stunning.

Now look more closely at the two images of M51 itself, taken at my lowest power, and then at the second one, which was imaged at a much larger scale than the first. The two images here provide a good comparison for the pros and cons of low versus higher powers resulting from image enhancement, and quite typical for many galactic subjects. We clearly can gain from both views, as long as the subject is sufficiently bright so as to allow higher powers. On the lowest power, the image is stunningly shown, almost bright! The arms are magnificent, even showing variations in brightness and texture along their length. The largest image is less bright, and the galaxy loses certain features so clear in the low power view. However, it does gain certain structural subtleties and considerable dramatic effect, not only because of the sheer size of the legendary "whirlpool" in the field of view.

I think M51 has become my all-time favorite, and still marvel at the impact it creates, with spiral arms swirling all around in the field of view, and quite brightly at that! In view of its visual prominence, it seems fitting, and indeed, most logical, that this galaxy would have been the one to first let humans see better into the fabric of the cosmos.

We now have more than just a taste of what is in store, and the Whirlpool serves as an indicator of the potential that enhanced viewing holds for us in dark skies. Another example is the grand and spectacular NGC 253, a slightly barred, almost edge-on SABc galaxy in Sculptor (Figure 6.3). It makes for supreme enhanced viewing, being both bright and relatively close. Missed by Messier due to its southerly latitude, it is a wonderful subject, hardly rivaled among any available to us. You will be instantly struck by its large size in the field of view, and it reveals all manner of detail, mottling and dark lanes in its central region, again, differently emphasized and much more brilliant when compared to the conventional view. The spiral structure is easy to appreciate, the primary arms appearing symmetrically from either side, becoming all the more defined in the image intensifier's view. In such a view, these arms appear more like short antennae than any traditional impression of broad spiral arms you may have. Part of this effect is due to the galaxy's orientation to us, and also to the nature of this particular structure itself. However, if you have ever seen an infrared image of this particular galaxy, you will be struck by how much the Collins image intensifier's view corresponds to such imaging. The reason is clear: with NGC 253, a third-generation image intensifier's strong sensitivity to such infrared wavelengths provides no better illustration of its best frequency response.

The full width of this galaxy, and the true dimensions of the halo as seen in conventional viewing, are, however, substantially less apparent now. For a very comprehensive sense of the whole, view NGC 253 through a regular eyepiece,

Figure 6.3.
NGC 253.

followed by an enhanced view. You will be left with a new appreciation of this old favorite, certainly one of the most magnificent sights in the sky. From a dark sky site, NGC 253 is so striking under any form of viewing, however, that I would hasten to add that there are many other examples of galaxies that present even greater contrasts between the different viewing approaches. However, take the time to marvel here at what is possible with your image enhancing equipment. The illustration reproduced here, a single frame, is, as always, taken from real time using my image intensifier. It defines for me, if asked to provide a good example for a novice to readily appreciate, what a stereotypical galaxy might look like through the telescope. That a true real-time image of this great galactic structure is able to give us another such representation, in many ways even bordering on the famous observatory portraits of which we have long been familiar, speaks volumes to the effectiveness of image intensifiers and enhancing devices in general.

In my quest for seeing spiral detail in a spectacular fashion, you have already seen that some examples of galaxies actually oblige us quite readily. M33, the famed Pinwheel Galaxy in Triangulum, is yet one more (Figure 6.4). This face-on Scd galaxy is often derided as completely invisible by inexperienced observers, usually because of its great apparent dimensions, and therefore its low total surface brightness. The next closest spiral to the Milky Way system after M31, I have actually had some success with it even from the suburbs using conventional viewing. With really transparent skies and lots of patience, I had often just made out its spiral form using an Orion Ultrablock filter, though never seeing more than just the faintest trace outline of the main spiral S-shape with this approach.

My Collins image intensifier never proved successful with this subject in suburban surroundings, and with its use any hint of the galaxy disappeared completely from view. However, I was already aware of the well-known extensive regions of ionized gas within the arms themselves, and speculated that M33 just might be a good candidate for intensification under dark skies, especially since it is a late type galaxy. Such structure as my narrowband filter had already shown was presumably due to these emissions. The potential for disappointment remained ever

Figure 6.4. M33 The Pinwheel Galaxy.

present in my mind, however. Maybe M33 might still be too spread out and diffuse to prove effective, even under dark skies, since most of M33's attributes do not fall exactly into the pocket of any enhancing devices' strongest features. However, to my great delight, M33 was surprisingly cooperative, giving a view that is just as striking as any type of live view of it I have ever seen. The "eye" of the red- and infrared-sensitive image intensifier gives new emphasis to the most well-known attributes of this classic.

The famous S-shape of the two primary arms, as well as other parts of the spiral structure, is at once apparent in the intensifier's field, and it also shows readily on real-time video. The main northern arm shows clearly, as does the famed huge glowing gas cloud and star factory, NGC 604, near its tip. Less apparent is the full extent of the main southern arm, and how it curls well back around the whole; it is faint, but careful examination will reveal it, even in the still frame here. The video frame certainly seems more illuminating than some drawings made by skilled observers in the 1970s, using professional observatory telescopes much bigger than mine. The detail, like so much in deep space, does not exactly jump out and greet you; but it is there nevertheless, and amazingly revealed at that. Also, we can plainly see stellar and gaseous concentrations all through the halo, and many more faint foreground stars than were obvious before. Be sure to compare what you see against a good photograph. Most of the subject is there, actually quite finely resolved! However, averted vision is particularly beneficial here, even when looking at the illustration. Appearing almost as in a photograph, albeit very significantly fainter, even novices can expect to see and enjoy it now with enhanced viewing. However, as an extended and faint object, it requires patience and persistence to fully take it all in.

Another amazing sight is NGC 4258 in Canes Venatici, later added posthumously to the Messier catalog as M106 (Figure 6.5). Impressive in any form of viewing from dark sky sites, this partially edge-on SABbc system nevertheless has some wonderful surprises in store for us with our new electronic tools. Usually described as bright, but not the subject of praise for the detail it shows us in conventional viewing, the spiral form is at once apparent now, as well as a general

Figure 6.5.
NGC 4258/M106.

unevenness of the general halo, which reveals darker regions between the arms. Note the ease with which you can see the arms swirling clockwise around the core, and faint extensions of the northern arm curving around, swinging sharply to the south. Since in conventional viewing the spiral form is only vaguely observable, and with averted vision at that, what becomes possible to see in this particular galaxy is another good example of utilizing our new approach. It is one of the most beautiful and awesome galactic sights, its scale and relative brightness assisting our viewing all the more. Imaged at a larger scale than many other galaxies illustrated here, it is one of the grandest sights in the sky.

A difficult object for many observers is the famous and grand SABc galaxy, M83 in Hydra (Figure 6.6). As one of the brightest late type spiral galaxies in the sky, one would expect it to jump out at us, even from the suburbs. However, with conventional viewing it is not only challenging in these less favorable locations, but is still not quite what one would expect even under dark skies. With enhanced viewing, though, we can finally enjoy quite a lot of success with it, and my first really satisfactory live impressions of M83 were with image intensifier.

Look carefully at the image here; the bar running east and west is immediately clear, as well as the spiral arms. The largest of these, the arm swinging prominently all the way across the northern side, then up and around the entire south side, it seems to follow just below a bright curved line of stars; they serve as a good visual reference to tracing the full extent of this northern arm. The arm to the south is much shorter and less pronounced, actually giving the galaxy a lopsided appearance. Interestingly enough, it is also possible to see a third spiral arm, much shorter and fainter than the two primary arms, on the far eastern side. M83 is full of dusty regions interspersed between the arms, serving to separate the spiral features.

Other objects that are just a disappointing smudge in less favorable skies suddenly come into their own in dark surroundings, often when we are least expecting such things. Experience has taught me, however, to sort through the most likely candidates first, but in spite of this, one never knows for sure what may work particularly well with any subject. A big surprise was the intensified view of

Figure 6.6. M83.

Figure 6.7.
NGC 2903.

NGC 2903, a many-armed Sbc spiral with a weak bar (Figure 6.7). It is well known as a beautiful sight in larger apertures, but nevertheless not usually described as one showing much spiral arm detail. Because of this, and its placement being almost face-on, it would seem an unlikely candidate for any such revelations in the enhanced view either. However, since it is bordering on a late type of galaxy, I knew I was at least in the running to see something. As it turned out, it obliged me readily, the bar showing very clearly, with the arms and "tributaries" jumping out at first glance, granting an exquisite view. The upper, bright portion of the bar, slightly separated from the rest, was apparently the subject of another classified identification by Herschel, NGC 2905, though it is only a brighter portion of the whole. Actual live viewing is even more revealing than any real-time video frame. However, in the illustration here, one is left with a vivid impression of the spiral nature and structural layout of the galaxy, clearly showing even on the printed image, another big winner for the new technology we are employing.

A different kind of galactic spectacle, and another wonderful sight for us, is the celebrated Black Eye Galaxy M64 (Figure 6.8). A bright Sab-type, and impressive to view with enhancing devices even in the suburbs, it is oriented not quite face-on to us. Its claim to fame is due to the huge dust lane lying to the northern side, apparently not echoed to the south. It would seem a unique phenomenon in the amateur's visible universe. This, the "black eye" itself, shows itself strikingly and darkly with enhanced viewing. (This feature can also be detected in conventional viewing at higher powers, though not nearly so readily apparent or contrasted with the main body of the galaxy itself.) On one side of the lane is part of a short spiral arm, and maybe hints of another larger one outside that, both of which just show up on the image here. The short arm is at the left side of the galaxy halo, and the suggestion of the other is under the black eye itself. The form of the galaxy is somewhat compact, with such arms as are visible wrapped tightly around the core. Even in time exposures you will not see very much else in the way of spiral structure. Because of the great contrasts within this structure, as well as its brightness, M64 will withstand considerable magnification, as is the case in the view presented here.

Figure 6.8. M64 The Black Eye Galaxy.

Now for one of the greatest surprises of them all: from my home location, M81, an Sab spiral, comes across as just another fairly bright fuzzy, and not at all impressive at that. I usually ignore its natural pairing with the irregular M82, when it comes to great visual treats. By comparison, under image intensification in such suburban locations, M82 alone can be spectacular indeed. However, in better surroundings, these galactic companions may again be seen among the most celebrated pairings of galaxies appearing often in the same low power field of view. What distinguishes them is not only their brilliance and proximity to us, but their utterly different manifestations of form and appearance. (See Figure 6.9.)

Placed in space at about 5 million light years from us, M81 is only twice as distant as is the Great Galaxy in Andromeda M31, and therefore is an obvious candidate for examination. However, the spiral arms of this famous galaxy are well known for their delicacy and relative faintness, so I did not anticipate the view that is actually possible, and with enhancing devices in particular. Amazingly, the chance of actually seeing something of M81's spiral structure

Figure 6.9. M81.

becomes a reality from a dark sky site, even in conventional viewing. Thus the galactic pair may be studied at the same time, and with the same level of interest and anticipation. However, when viewing M81, we must still be patient and critically observant, no matter how we approach it. It remains a difficult object, but with care and practice, it is indeed possible to begin to make out something of a spiral arm or two in conventional viewing.

Surprisingly for an Sab spiral, and much to our good fortune, I have found that wonderful things also begin to happen when we turn our enhancing gear on it as well. Originally, while figuring I had nothing to lose, I turned my I_3 on M81, assuming I would quickly move on to better things. I could not have been more wrong. Look carefully at the image here, which is, as always, just one image intensified video frame, taken as always from real-time video footage. It seems quite miraculous that you can clearly see the delicate spiral arms swirling around the galactic core, although they are indeed subtle features. Hopefully you will forgive the numerous bright triangular foreground stars, overexposed during my imaging, to allow M81's faint wonders to shine through as they appeared to my eyes, live.

(The other galaxy in this grand pair, M82, is featured in the section on irregular and other unusual galaxies.)

Similarly, I never expected much with M61 (NGC 4303) in Virgo (Figure 6.10). This is a true face-on system, dramatic in photographs, but sufficiently interesting to me in appearance that I considered it likely to let me down! Documented images you may be familiar with show a strong spiral form, though with one of the two main arms exhibiting a strange triangular hook in shape, something like a coat hanger. Remarkably, here was another great triumph for my image intensifier. In that this is a galaxy rich in blue light, the results speak for themselves, once again delivering more than the device seems to promise. It not only showed the spiral structure fairly brightly and readily, but the famous hooked arm as well, which in real-time seems almost ridiculous! The video frame here gives a good approximation of its visual appearance, but nothing quite makes the statement like seeing it for yourself. Note the two 14th magnitude stars, conspicuous

Figure 6.10. M61.

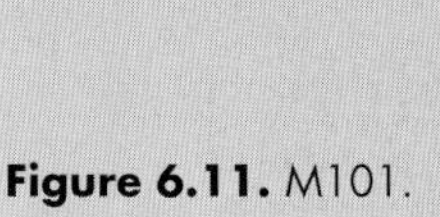
Figure 6.11. M101.

on this image; other fainter stars can just be detected if you know exactly where to look. Interestingly enough, there are the beginnings of a third spiral arm also just visible here, pointing almost straight up near the top of the image of the galaxy. Time exposures show this as a rounded and blunted nub, and it certainly has never developed to a degree approximating the other two.

Let us now travel to M101 (Figure 6.11). Here is a late type SAB galaxy, wide open to our viewpoint, with arms spreading out over a considerable area in the field of view. Again, as with all truly face-on systems, I was always prepared for a letdown. However, this particular galaxy is well known for its numerous HII regions, splotched throughout its arms and well out into their far-flung extremities. Because the features are subtle, even in conventional viewing, you will need to study carefully what you are seeing.

Look at the illustration here alongside a good professional observatory or CCD portrait. With enhancement, virtually all of the galaxy and its HII regions are visible in our real-time view, and you may ultimately count the sight through the telescope as one of your most significant views of any galaxy. Do not expect it to be bright, though, and you will need to look at it carefully. The most prominent spiral arm, to the north, shows much mottled texture, and continues around the core all the way to the eastern side. Just below the brightest section of this arm, nearest the core, is the well-known dark region, in some ways similar to the "black eye" of M64, though less pronounced. Here, at the northern end of the dark region, the arm splits into two segments, the outer portion continuing outward significantly further from the central mass, almost completing the circle. The eastern arm curls around the galactic nucleus in a comparable fashion, and contains many HII regions, visible here as faint irregularities and bright spots. Near its origin, this arm, too, divides into an inner and an outer arc, the outer segment extending far from the galactic nucleus. Be sure to trace the full extent of the arms, as well as the faint but present nebulous HII regions placed far away from the center, although they may not be readily apparent without a good reference image. I found it necessary to stretch the contrast considerably in the video frame in order to give a decent representation of the presence of the live view; doing so does degrade the image to some degree.

Figure 6.12.
NGC 3726.

Tiny NGC 3726 in Ursa Major is another face-on galaxy that holds a surprise for us in the viewing of spiral structure (Figure 6.12). Although a late type SABc, its small size would seem to preclude it from yielding much detail. Typically it is described by observers with fairly sizable instruments as showing only hints or traces of spiral structure. Again, because of the presence of desirable HII regions in the southern arm, I thought it worth a go, however. This feature, the larger of the galaxy's twin arms, seemed eager to disclose its being to my I_3, although its presence is less striking than the northern arm, which jumps out, wrapping prominently around the core in a tightly wound corkscrew. Hardly a struggle to see by comparison, the southern arm has a bright blob of a knot along its length, which makes this entire sight look not unlike a diminutive Whirlpool Galaxy. The knot is presumably a vast region of extremely ionized gas, huge by any standards. Once again, we do not normally expect to see such clear indications of spiral form in such a diminutive face-on galaxy with any type of real-time viewing, so the strength of my equipment once again came into its own. The delicacy of this particular spiral is striking, and its slight barred structure is also just apparent.

We should comment at this juncture on the famous pair of spiral galaxies in Leo, M65 and M66 (Figures 6.13 and 6.14). (They are associated with NGC 3628, a grand edge-on system, which is featured in the section on edge-on galaxies.) For revealing detail, the two are worthy sights, at least photographically. However, it is uncommon for the live observer to see much in the way of spiral detail in either, though larger apertures reveal much mottling and at least hints of the galaxies' structure. Using an image intensifier, once again we are treated to some more of the visual puzzle and can begin to make out more of the actual forms of these galaxies. Their outlines become clearer, and armed with a little knowledge of their forms, what we may see will be actually quite complete, in as far as the main elements of the structures are concerned.

In M65, an SAB galaxy, we can plainly make out the notch-like, darker spiral regions of its south and north limbs, on opposite sides of the core. Considerable mottling and unevenness in the general halo are also quite apparent, as is the overall outline of its shape. It is not, unfortunately, the easiest subject to tackle, and the long dust lane traveling all along its western edge, easy to spot on

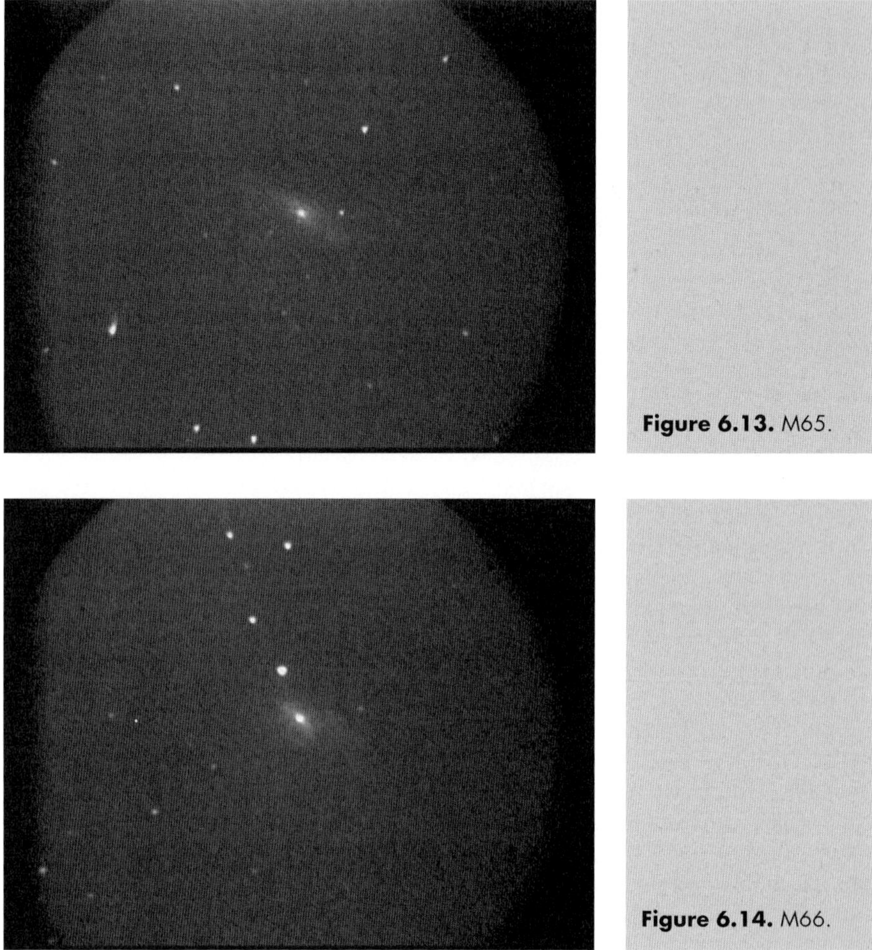

Figure 6.13. M65.

Figure 6.14. M66.

photographs, seems elusive. Part of the problem is that the lane lies adjacent to a very faint and narrow arm, which provides us with little to contrast it against. However, the small stellar point, east of the core, is superimposed on the lane itself, so, for traces of the lane, you will know where to look.

For seeing something of the spiral arms, M66, a SABb galaxy is a different matter. Being quite lopsided in the view it presents us, the long and wide southern arm trails off into space, while the northern arm hooks around the core, presenting a live view altogether in line with its more well known photographic portraits. It is generally surmised that the distortion was caused by a tidal interaction with the grand and explosive galaxy, NGC 3628, in the past, which drew the southern arm outward into this curved stream. Certainly it is unusual for amateurs to describe seeing any spiral pattern in M66, but my intensifier reveals it unmistakably, and careful inspection leaves little doubt as to what we are seeing. Even the slightly barred central region appears visually as a stronger feature than would be expected by its designation.

Figure 6.15.
NGC 5248.

The potential for great sights among spirals goes on and on; do not think for a second that the list is near to being exhausted! Some others are briefly included here, just so you will have even a clearer idea of our potential, especially with subjects not exactly well known. The somewhat diminutive spiral, NGC 5248 in Pegasus (Figure 6.15), shows well: two nearly equal arms jut out from opposite sides of the galaxy. An especially bright region, looking almost like a star, is present to the east, as well as a small dark band, almost due south.

Also in Pegasus is the wonderful, partially obliquely oriented SAb galaxy NGC 7331 (Figure 6.16). Having dusty regions encircling it, traces of spiral structure can be seen among them, extending far outward from the immediately visible bright central region, clearly visible as darker gaps in the halo around the north side. Several fainter, small associated galaxies are also easy enough to spot with our powerful light-amplifying tools. The detail is limited, however, to no more than this, but the entire view presented is nevertheless quite beautiful.

Figure 6.16.
NGC 7331.

Figure 6.17. NGC 7479.

Another famous but normally quite tricky galaxy in Pegasus is the wonderful NGC 7479 (Figure 6.17). Difficult as a real-time object in any circumstances, it nevertheless presents us with some real potential away from the city. With enhancement, the bright core and, more especially, its famous S-shaped form is relatively easy to see, though I should caution you against expecting a dazzling show. I find it extremely gratifying to finally see this familiar form, live, through the eyepiece of my image intensifier, altogether more clearly revealed than any other way I have seen it. Use indirect vision, even when viewing this image. Note the significantly brighter southern arm, swinging westward, enveloping one star and seeming to touch another at its farthest point. The northern arm appears more nebulous, just missing a bright stellar point at the extreme north end of the galaxy.

A very pretty galaxy, NGC 4725 in Coma Berenices, is also worth a shot (Figure 6.18). It may be too faint for many telescopes, but if you are successful, you will become at least partially aware of some its most striking features: a bar, at some-

Figure 6.18. NGC 4725.

what of a slant in reference to us, the arms that seem to swirl around it in an oval, together with many small foreground stars sparkling within its outlines. Though a difficult galaxy to see live, the sheer beauty of NGC 4725's form warrants the effort you may make; maybe it will reward you, although the image included here shows how challenging it is to catch. By all means prepare yourself ahead of time with good knowledge of its appearance in time exposures. Again, spiral detail is clear, spotted with stellar and gaseous concentrations along one side.

Edge-on and Nearly Edge-on Galaxies

Some of these subjects are among our best subjects to view live. When they are girdled by dust lanes, their double-sided profiles are memorable indeed. Among edge-on systems (a Generation III image intensifier's best-suited galaxy type), NGC 4565, an Sb galaxy in Coma Berenices, must surely represent the ultimate example to everyone who has ever seen it (Figure 6.19). However, only when seen in truly dark skies will the differences between it and all other edge-on systems be fully appreciated; nothing else even comes close.

The enhanced live view of NGC 4565 probably best approximates all of the grand photographic portraits of edge-on galaxies we have become accustomed to seeing. The famous dust lane is a wonderful sight: wide, very dark, and reaching out into the surrounding space. It is this particular feature, of course, coupled with the galaxy's large size and prominence in the field of view, which sets this galaxy apart from all of the others. Use averted vision to see it best, even when examining the illustration here; you may even be able to detect some irregularities along each side of the belt.

The plane of the galaxy, just slightly tilted relative to us, presents the core just below the belt, glowing as a bright star-like point (almost appearing like the pupil of a sleepy eye; one commentator reflected that it seems to wink at him!). This

Figure 6.19. NGC 4565.

core region bulges conspicuously, visible just as in photographs (not unlike two "hi-hat" cymbals on a stand together), and slight mottling along the almost perfectly symmetrical length can be seen. Magnificent! In most other edge-on galaxies such shapely bulging cores are not usually nearly so strikingly contrasted against the rest of their needle-like structures. Almost needless to say, the image here reproduces only a fraction of the impact of the stunning real-time view, or that of actually seeing this awesome galaxy's eerie gaze for oneself.

NGC 891, another edge-on Sb galaxy, this time in Andromeda, is almost as imposing and well known as our winking friend, NGC 4565. Not nearly so bright as its more famous cousin, it is hardly visible in the suburbs, though; even using moderate and large apertures, it remains difficult to locate. It took me many observing sessions from my home location to actually be sure I had seen it. Even now that I know what to look for, on many occasions I am unable to detect it at all from my home site, no matter what I do, although it is not especially faint in documented magnitude, compared to some other galaxies that show more readily. Some of this apparent faintness is due to its relatively large angular size in the field of view. (See Figures 6.20a and b.)

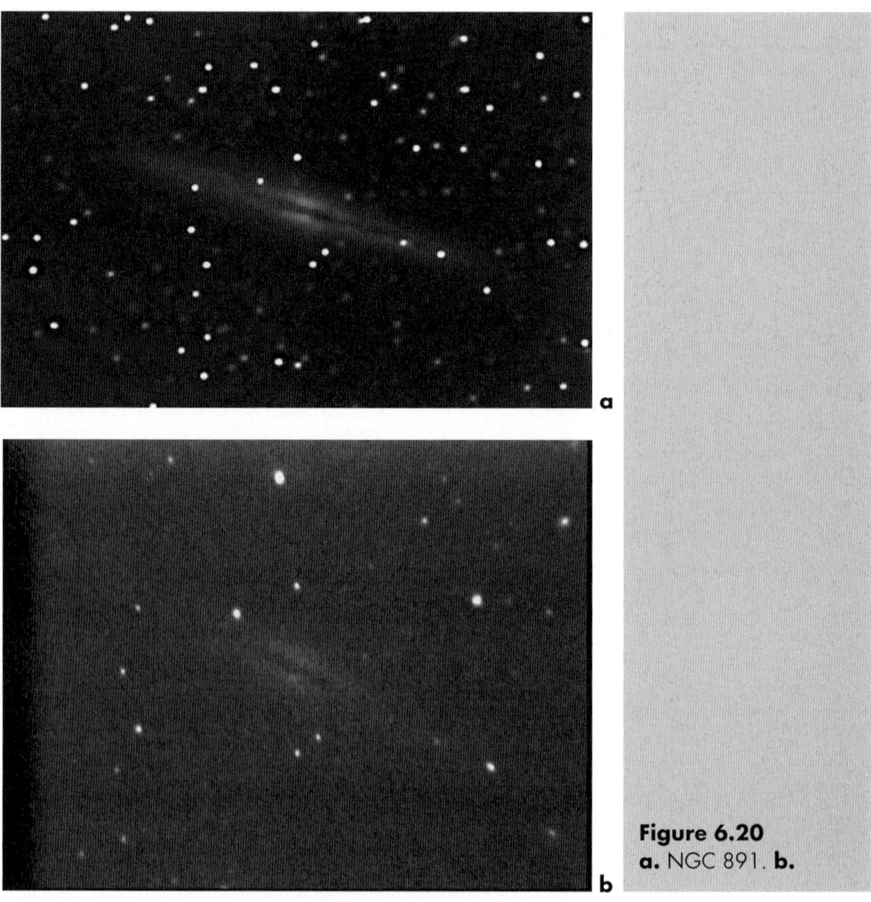

Figure 6.20 **a.** NGC 891. **b.**

Now under dark skies, once located, it is hardly less impressive than NGC 4565 in its total effect, although it is not as bright, and is less defined in shape and contrast. Amazingly, it is at once easy to see well with conventional viewing, even in moderate apertures, and the beautiful star field setting in which it floats appears almost like a cosmic jewel box. There are few settings more magical. What is most striking, though, once we turn our image enhancing systems on it, is the stunning dark band and mottling visible along the galaxy's entire length. Suddenly, NGC 891 appears in essence as we might expect from photographs, and also shows up well on real-time video. With conventional viewing, even from these desirable locations, this galaxy's special features, namely its dust belt and other irregularities, are still difficult to see without some effort, because of the galaxy's low surface brightness compared to its size. Detail is extremely subtle. The views of NGC 891, reproduced here, are fairly representative of its appearance with enhanced viewing, an altogether different story. The first (Figure 6.20a) shows the exquisitely smooth and telling still images that may be extracted from using a StellaCam EX integrated video system (image courtesy James L. Ferreira and Adirondack Video Astronomy).

Most striking here, of course, is the famous central dark band, now quite easy to see. Again, using averted vision, note the mottled extensions of the galaxy, and even a central bulge of sorts. You can begin to see why it is spoken of frequently in the same breath as NGC 4565. Initially, under intensification, what struck me most about this particular spectacle were two things:

1. Its easy visibility, compared to the strained view from the suburbs.
2. Its similarity in brightness and general appearance, as seen under dark skies (aside from its being immersed in a rich star field) to NGC 4565 as seen from the suburbs. These relative views tell you something of the even more stunning view of the great NGC 4565 that lies in store for you from dark sky sites!

There are few galaxies that are so strongly suggestive of a star system far across space as is NGC 891, and although there are many wonderful edge-on systems that feature dust belts prominently, few will appear more memorable. This is the unique quality of NGC 891; among true edge-on systems, I would rate it third, only to NGC 4565 and M104, the next galaxy we shall stop by!

Very close to the description of edge-on, and as spectacular as any in the sky, would be M104, the famed Sombrero Galaxy (Figure 6.21). It is quite successfully seen even with conventional viewing, but nevertheless it withholds its greatest impact for the enhanced approach, with which it puts on a very good show even in suburban environments. It is noticeable that in true darkness its dimensions are further extended, and it reveals its celebrated shape and huge dusty girdle even more strikingly. Do not expect to see any evidence of spiral structure, however, since these are very dense and finely resolved features visible only on time exposures. Fear not; you will not be disappointed with what you see. Here is a near-edge-on system that bulges very obligingly for us, so much so that the distinctive sombrero shape resulting, and which inspired its name, is firmly established in our eyes and imaginations. This is the only galaxy we can access in real time that shows a profile such as this: its famous hat and rim form, quite obvious in the field of view.

Figure 6.21. The Sombrero Galaxy M104.

The massive and dark dust belt, a defining feature of this very symmetrical galaxy in itself (the hat's brim), is among the widest and most defined of any available to us in real time, indeed comparable to that of NGC 4565. However, its sweep upward at each side is seemingly quite unique in the amateur's universe, owing its appearance to a less than perfect edge-on positioning of the galactic plane relative to us, as this wide and dark lane sweeps around to the rear of the galaxy. The prominence of this feature makes it really noticeable at first glance. The blazing core at the center of the galaxy resembles a huge lightbulb, seemingly lighting up the surrounding space, as the galaxy hangs in space amid the stars, like a grand decorative chandelier at a fancy ball. M104 remains one of the greatest galactic showpieces of them all. Look at the image here indirectly in order to appreciate the celebrated dish-like form.

From one extreme to another when it comes to edge-on systems, the lenticular system NGC 5866 shows a dust belt about as finely defined as any we could hope to see (Figure 6.22).

Figure 6.22. NGC 5866.

NGC 5866 is a definite *wow!* One of the surprises of suburban viewing, I have always found NGC 5866 to be a very friendly galaxy, ready to give up its secrets when our new toys are put to work, as long as we use adequate magnification. Moderate apertures should reveal it as an unusually featured galaxy, with quite the finest dust lane of any edge-on system I can recall ever seeing. Indeed, in *The Night Sky Observer's Guide*, only slight indication of the dust lane is clearly noted with apertures much under 20 inch to 22 inch (51 to 56 cm), so one would never expect to actually see it easily in real time with anything less. However, with my 18 inch, under the poor conditions usually experienced in the suburbs, I saw it easily the first time out when using my image intensifier!

This narrow lane should even become a fairly easy mark in telescopes half the size of mine, equipped with enhancing devices at dark sky sites. If you look at any good professional portrait of this galaxy, you will be struck by the tiny and refined appearance of the dust belt. The neat and almost exact dissection of the galaxy into two equal halves is actually more unusual than it seems it would be, among edge-on systems. Because of the unexpected and remarkable appearance of this fine lane, I have always rated NGC 5866 as one of the most satisfyingly resolved galactic sights we have. Look for the fine stellar-like point at the end of the galaxy's more southerly extension, and the field of stars that frame the scene. Certainly the magnitude usually assigned to the galaxy would seem to suggest it would be less striking than it is in the field of view. However, its prominence has prompted some observers to dub it M102, not because NGC 5866 is actually the missing Messier object, or in the correct location, but because it fills in the blank designation within the catalog so convincingly!

Something like NGC 4565 in appearance, though considerably fainter, is NGC 5746 in Virgo, slightly south of a seemingly dazzling 7th magnitude star (Figure 6.23). Apparently an Sab-type spiral, an encircling dust lane may be clearly seen in the live enhanced view, on the northern side of the central core. Near the tip of the south extension are two bright points, as well as some other tiny stellar-like features near the northern extension, very striking by the usual standards of what we can see in such galaxies. From this side perspective, it looks like a

Figure 6.23. NGC 5746.

Figure 6.24. NGC 5907.

plano-convex lens, more so than other edge-on systems I know. Do not expect it to rival the best and brightest examples of such systems, though it puts on a pretty good show. It is likely you will be considerably more impressed with the live view than you will be by the recorded image I was able to extract here.

In some ways, yet another near double of NGC 4565, right down to the prominent star adjacent to the nucleus, is the Sac-type galaxy, NGC 5907 in Draco (Figure 6.24). The similarities end with the general resemblance, because in total illumination NGC 5907 is hardly a contender. Its overall outline is also substantially flatter than NGC 4565, and without the striking central bulge. The dust lane, while visible, is not nearly so immediately prominent, especially since it is less centrally placed, one side dominating the view with most of the illumination. In the image here, it is possible to discern the lane (on the side with the prominent star), but it is easier to see with averted vision.

In angular size, NGC 5907 is only about half that of its more famous relative, but nevertheless it is quite a wonderful sight in a telescope of sufficient aperture, as it eerily peeks out at us from far away. Careful inspection of the galaxy will reveal some mottled texture. Without image enhancement, this galaxy is likely to be a disappointment to those becoming accustomed to brighter and more spectacular images.

Quite a famous galaxy, also with a prominent dust lane, is NGC 3628 in Leo (Figure 6.25). Depending on the magnification you employ, this splendid Sb peculiar galaxy may also be in the field of view when you view its nearby and brighter galactic neighbors, M65 and M66. Classified as a peculiar galaxy because of its explosive evolution, it is spewing newly created stars into its entire bright central region.

Well-known observatory images show NGC 3628 as a grand spectacle, more imposing in these views even than its more celebrated neighbors. Unfortunately, in live viewing, it is also considerably dimmer, and within city limits the galaxy is usually invisible, or vague at best. However, it is well worth the trouble to seek out in a dark sky; from the desert it proved an easy sight. Still not as striking a spectacle as some other more famous edge-on galaxies, it was nevertheless a thrill

Figure 6.25.
NGC 3628.

to finally see it, and to see it quite well, although the galaxy is much brighter in the bulging central region than elsewhere within its span. The dust lane stands out prominently, and seems to widen and become more diffuse along its length. In the enhanced view NGC 3628 is narrowest near the core, its slightly twisted shape apparent under careful examination. If this large galaxy had only been brighter, there is no doubt that it would occupy center stage in the neighborhood.

Another of the great surprises under the cover of dark skies was the sight of NGC 4631, an SBd spiral in Canes Venatici (Figure 6.26). Never satisfactory in the suburbs, it makes a major statement in our present situation. Appearing more like an irregular galaxy, it is loaded with mottled detail, and has substantial dimensions in the field of view. Its shape is very tapered and fairly irregular, with faint detail trailing off into the reaches of space. Near the center is a bright point, in fact a 12th magnitude star. Interestingly enough, there is not a trace of a dust lane surrounding this edge-on system, although it is truly wonderful to view.

Figure 6.26.
NGC 4631.

Figure 6.27. NGC 7814.

Finally, among edge-on systems, perhaps the most remarkable demonstration of the effectiveness of image intensification (though anything but "stellar" in terms of beauty of image!) would be NGC 7814 in Pegasus (Figure 6.27). I cannot tell you that the image included here will make your heart pound. Visually, it is a stretch, but I have included it here since it is such a telling example of the power that image enhancing devices have to unlock the universe in real time. You may be aware of the celebrated appearance of this galaxy in observatory, or fine CCD, images; the well-known dust belt is completely obliterated by the central blaze, appearing like a welding arc; it appears to have been burned out of existence at the galaxy's waist. In conventional viewing, most observers can only dismiss it as just another elongated blob, and the galaxy itself is certainly not bright enough to be glimpsed at all with regularity from the suburbs. Under image intensification at our remote location, it might just reveal its real claim to fame, given patience and care. Suddenly, the fine, but otherwise usually unseen dust belt (through amateur instruments), jumps out from time to time. The "welding arc" does indeed completely obliterate the dust belt here, as in all of its well-known portraits.

Irregular, Elliptical, and Unusual Galaxies

Certain galactic structures other than spirals are outstanding subjects for us, and are sometimes of highly unusual appearance. (Some of these have already been covered in the previous section, because they may well be edge-on to us, and present striking viewing in their own right. Although edge-on galaxies are not a galaxy type, they nevertheless present themselves to us in ways that make them appear so.)

Perhaps the brightest and most immediately striking irregular galaxy is M82, the other galaxy of the famous galactic pair (with M81) in Leo (Figures 6.28a and b). M82, a somewhat bizarre and seemingly explosive galaxy, radiates a wealth of

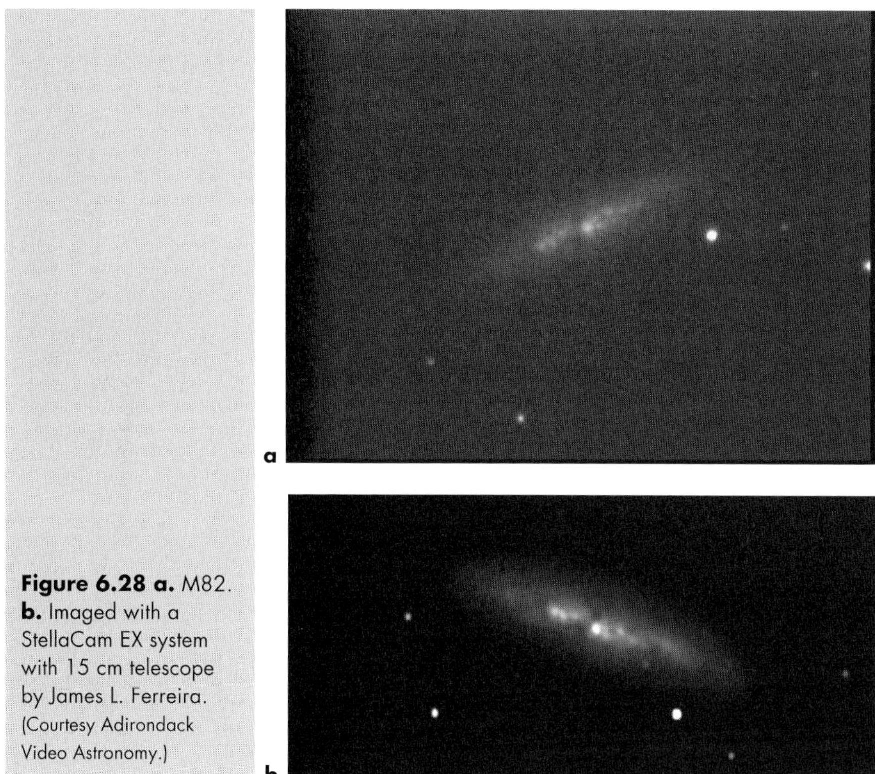

Figure 6.28 a. M82. **b.** Imaged with a StellaCam EX system with 15 cm telescope by James L. Ferreira. (Courtesy Adirondack Video Astronomy.)

light wavelengths ideally suited to our purposes. Indeed, many irregular and peculiar galaxies present us with some great spectacles. M82, in my view, is nevertheless the king of them all.

Even from the suburbs, M82 is capable of delivering startling results with enhanced viewing, yet comes even more into its own under the canopy of the dark sky. It could not be more perfectly contrasted with its beautiful and perfectly formed spiral companion, M81. It is not that so much more can be seen of M82 now in our truly dark surroundings, but the added brightness of the detail present makes what was usually a suburban viewing rarity into a common and even better displayed spectacle. The image here reveals no more than would the live intensified view with just a moderate aperture in dark and fairly transparent conditions. Note the incredible mottling, dark veins, and lanes that are so visible even here. The impact of the view through an aperture such as through my own telescope will make your jaw drop! Mine does, every time. Under dark skies, the sensational pairing of M81 with M82, seen easily in the same field with a low power, wide-angle eyepiece, becomes a more exciting visual option than it ever was in the city.

And finally, one of my personal favorites among all galaxies, the stunningly bizarre NGC 5128 is another highly unusual galactic structure (Figure 6.29). Unique in the amateur's visible universe, and classified as a peculiar S0 type, it certainly fits the description of a peculiar galaxy, appearing much as a round

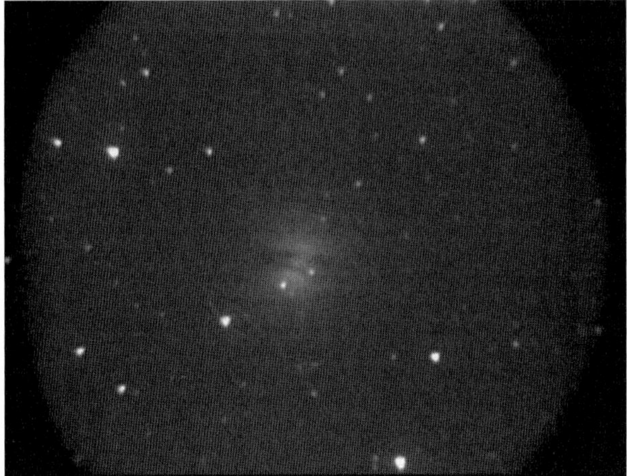

Figure 6.29. NGC 5128.

lenticular system with a huge dust girdle swirling around its midsection. Apparently, this gigantic galactic anomaly, a powerful radio source, is actually be the result of a collision of two galactic systems, an elliptical and a smaller spiral. The dark banding would appear to be mostly the remnants of the smaller spiral galaxy – the one that lost the encounter! Over time, it will fully merge with the huge parent elliptical to become an even larger, but totally amorphous sphere, an end product that apparently lies in the future of all galaxies.

Even from the suburbs it is a worthy spectacle for enhanced viewing under the right conditions; but here, under dark skies, it blows me away! Most striking about this galaxy, using enhanced viewing in dark skies, is its totally dazzling visual presence, compared to the much fainter apparition under less favorable conditions. It almost leaps out of the telescope! This is not likely to be the case when using conventional viewing, and as such, NGC 5128 is not often described as one of the prime galaxies to view. But now, looking carefully, you will be able to detect the considerable girth of the halo of the galaxy, a far wider structure than generally realized in other types of live viewing.

Note the dark banding, which shows something of its great width in the image here, much greater than the brightly illuminated part of the galactic halo; when viewing it, this seems darker than any of the immediate surrounding space. Slicing the galaxy core into two halves, the southern side is noticeably more prominent with several stars superimposed on it. Note that even the band is again divided into two halves, the inner divider glowing like islands with the fire of cosmic conflict, culminating in a bright point on its eastern end. This detail within the banding is brightly shown, as is also the extent and twisted shape of its entire form. This particular subject's lowness in the northern sky will render it out of range for many observers; more the shame, since it remains one of the most startling and mystic sights in the heavens. It still seems amazing to me that it is possible to see it with such brilliance and detail when it hangs so low in my northern sky.

You will find many other worthy sights to look at beyond those listed here. In no way is the survey presented in these pages supposed to represent more than a suggestion of all that is possible. You will need to explore all that you can under your own steam (forgive the nineteenth-century phrase; somehow it seems appropriate for the visual approach!), and armed with the new means to achieve our ends you are not likely to be disappointed. I think the potential is clear and dramatic. But now, I would like to return to earth, and ponder a few things on the future of amateur astronomy. Hopefully, in your hands, it is in a good place.

CHAPTER SEVEN

Eyes on the Future

So here we are, at the end of the beginning of a new journey. I say this because it is my hope that it truly is the start of something new and exciting for you in what is actually quite an ancient arena. This, in a time when I just cannot help sensing that the old-fashioned art of real-time visual observing is under siege, but more on that momentarily. Visual astronomy has benefited many times in the past by new technological advances, so why not now? Therefore, it is my hope that this book will point you to ways to expand your potential, in what was already a satisfying and completely absorbing subject. Perhaps this will do something to keep the spirit of real-time visual astronomy alive. Maybe it will also point some newcomers in a direction other than what may seem an inevitable one these days.

Astronomy in any form is apparently an addictive pursuit. Why else would otherwise-normal people go without their needed hours of sleep, and often in such uncomfortable conditions? Maybe the answer lies within that very question: perhaps we are not so normal after all! So, however abnormal we might actually be, we find ourselves responding to that intangible force within all of us that leads us to look to the stars. After all, it has been said that humans differ from the animals in that occasionally we pause to look up at the sky.

Astronomy is something in which the mind and imagination can leave earthly bonds, and dare to envisage worlds other than our own. Can there be any other typically solitary pursuit in which so many hours pass as quickly? It seems no sooner have we set up for a long night of viewing that dawn is suddenly upon us again. It is already time to pack up and call it a night. If ever there was confirmation of our inability to comprehend the fourth dimension, then this is it; the time spent among the stars is apparently barely detectable by human means! And can there be anything else where the mind is kept more constantly energized by all that we have experienced, long after the fact? All of it remains with us forever, playing over in our minds and imagination, much like great music. Some say that music relates to the very fabric of the universe itself, an analogy that seems fitting. I am not the first to bring up the link.

Although, out of necessity, the bulk of our time under the stars remains a somewhat lonely occupation, it is likely you will often wish to share some of the wonders you have experienced with others. These times will not be when you conduct your most important and all-consuming efforts, such as drawing or imaging, but they will be some of the most memorable. In this alternative and more social role, dark sky sites have a splendid multiple billing. Because of the natural viewing attributes of such locations, they are the best places to share our space travels. Not only do we have companionship of a participating person or persons (and under some of the best of viewing circumstances), but we can share with them all that inspires us.

No one is likely to be disappointed out in the wild with the visual treats in store for them; the smallest of telescopes, even without enhancing devices, can open some jaded eyes under such skies. Friends can provide new incentives to explore, which will make the most marathon stargazing sessions stay infused with energy, instead of sometimes with weariness. Our friends can also share in the simple logistics, such as sharing the driving and setting up camp, things that wear out the hardiest among us. They provide an unspoken sense of reassurance against being overwhelmed by our often stark and lonely (dare I say, even hostile?) surroundings! Best of all is to have their eyes (often along with those of countless furry critters!) as our chosen space travelers to relish our discoveries. They make what would be otherwise a loner's pursuit into quite a social one. On a purely selfish note, they can help us in carrying and setup of our heavy equipment! (See Figure 7.1.)

The more we reach out, grow, and share what we have with others, the greater the likelihood is that we will actually recapture some of the innate, almost naive wonder many of us had in the past, the time when we first became fired up by astronomy. For me, these memories are forever and indelibly locked into my psyche. There is no doubt they were also colored by my youthful immersion in such texts as *Amateur Telescope Making* (Ingalls; *Scientific American*), with its

Figure 7.1. Preparing for an all-night viewing session in the desert with friends Andrew Shulman (left), Jim Corley (right).

inspiring debt to the great Russell Porter. This set of three, jammed volumes unquestionably has colored my whole approach to practical astronomy. With increasing technical advances, ever widening commercial availability of equipment, and the crossovers from other areas of expertise, some talented and dedicated individuals, such as telescope visionary Porter before them, have even managed to blur the lines where amateur pursuits end and professional activities begin. For those notable figures who did this in the past, it seems to me that it was a more difficult journey to successfully navigate during their time. We will always look up to all of these larger-than-life historical figures, including William Herschel (probably the most significant of all of them), Milton Humason (Hubble's assistant), and Clyde Tombaugh (discoverer of Pluto), all heroes of the golden age who went on to towering heights of accomplishment in astronomy. None of them had any formal training as astronomers.

Other textbooks also provided inspiration that was indelible on my impressionable youthful mind. What have now become historic manuals, such as *The Larousse Encyclopedia of Astronomy* (Rudaux and Vaucouleurs), imprinted a sense of wonder and awe for all of those earlier twentieth-century figures, even those of the nineteenth century. They also gave us an appreciation of the monumental new tools and constructions they used that led to modern astronomy and cosmology. Those historic times were indeed an incredible age of wonder and discovery, fueled by large helpings of imagination in all those who pursued the unknown. For all of us astronomically inclined, following as enthusiastic amateurs in their shadows, the stamp was lifelong, having been heavily influenced by books such as these. I do not think I would have had my interest today had I not had those associations from my earlier years to draw from. Many young people today will never know these inspirations, at least partly because of the popularization and ready commercialization of astronomy these days, with all of its trappings. Suffice it to say, for me, the dark skies only invigorate the possibilities laid down upon this foundation, which were created under far less favorable circumstances and opportunities, astronomically speaking. It is amazing just how many soul mates I have run into in campsites almost "next door" out in the great beyond, doing just the same thing as me. It seems we are all chasing the same dream.

Visual Astronomy Under Dark Skies completes the second of two books primarily concerned with a special and somewhat different approach to visual astronomy. I am all too aware that my particular pursuits are not in line with the direction in which many people are trying to push amateur astronomy these days. The great god of CCD imaging seems to be making ever-increasing inroads into the field, and although there is nothing wrong with that in itself, it seems many people are prepared to accept an inevitability that the CCD way will ultimately become the dominant, indeed even the only force in amateur astronomy! I have already made some mention of this, and the direction it seems we are going, in Chapter 1. Not to take anything away from the great god, of course, but have all of the older devotees of our wonderful hobby forgotten the visual quest that presumably fueled their own drive, once upon a time? By giving short shrift to the value and inspiration found in real-time observing, aided by some very real and relevant technical advances available to us today, these same people are almost ensuring that their own (pessimistic?) view of the inevitable takes place. This is instead of embracing and encouraging a world where both approaches can live together; indeed, one should feed the other. Must one path necessarily be at the expense of the other? Live viewing provides an almost spiritual bond to the cosmos. This is not to say that other types of bonds cannot also be experienced by those who take different paths. However, these paths do not converge, or even take us to the same place.

I also commented in Chapter 1 upon the great arrays of sophisticated equipment that are so widely in use today, including those for the amateur observer. Looking back over the near lifetime of my own involvement in astronomy (so far back, it is hard to remember exactly when it started), our present sophistication seems like a two-edged sword. We dreamed of having all that those things that today's sophistication and availability have brought us, and those dreams fueled our quest and enthusiasm to reach for them. Now that we have so much of what we dreamed of, a lot of us are not dreaming any more! I feel all too aware that where we stand today has in some way caused a lot of the magic and wonder of the earlier "golden age" of amateur astronomy to be lost, maybe forever for some. This genesis fired up the mind's creative processes, centering around the visual experience, and all that it took to get there.

It certainly cannot do any harm to reflect on where we stand today, in order that we can see from where we have come, and to serve as an indicator of where we might be going. I hope my thoughts will encourage others not to allow the magic and wonder of amateur astronomy's original mission to slip away from us; we should not allow it to disappear while we remain blissfully unaware. It is, of course, the same one that fired up all of those now-legendary Amateur Telescope Makers of Springfield. In spite of the efforts of these pioneers and others, along with their collective success in popularizing astronomy, I have the feeling that many of these remarkable pioneers might be less than thrilled with every aspect of amateur astronomy as it has evolved. There seems to be a major part of it that has become cold and detached, even routine. I hope it is not too late to change course for something that embraces more our original passion.

The present state of affairs partly in mind, I started to compile a list of the things that have changed so noticeably over the relatively short period of my lifetime. The list turned out to be larger than I had anticipated. Certainly the advances are astounding and significant at any level, but upsides also have downsides. Following this line of reasoning, I realized that the subject had become quite a sizable arena for discussion in itself, and so I discarded my list in favor of a much more far-reaching look at where we stand today, as well as where we are going. In the following critical review of this subject, I do want to stress ahead of time that I do not mean to advocate a return to days of old, just the preservation of the ideals, and mostly the spirit of those times. Of course we have to march on; there is, of course, considerable value to the things that have advanced our cause, and all that we have gained in convenience, or even take for granted. We can also take all available means to help us do so more productively. However, I continue to hope that the march onward will leave us better off and even more inspired, and not the other way around. Everything has its place; it is just a matter of ensuring that everything can find its place to everyone's best advantage, without diminishing (or even eliminating?) another part of the whole.

Perhaps the greatest single advance that we see today is the wide availability of highly sophisticated equipment in the marketplace. Along with this has come an amazing drop in relative costs, at least as far as the truly mass-produced lines are concerned. Also remarkable is the fact that the actual quality and consistency of this type of mass-produced equipment has drastically risen, to a degree that would have seemed impossible not so many years ago. Even all of the available accessories stand at a level and variety that was reserved only for the military in my early days. New technology has also brought down the prices of many of them, and we also live in a time that allows more discretionary spending for the average person. Modern astronomical equipment includes not only a wide array of telescopes and related products, but radical eyepiece designs, which have progressed to an unforeseen degree. These eyepieces have corrected many optical problems (except bad quality!), making short-focal-length telescopes stunning performers; once these scopes could handle very little power and had quite limited applications. Some considerable apertures now have become practical in the amateur community, along with refinements never dreamed of by our predecessors.

With some modern configurations of old designs, coupled with innovative solutions to old problems, has come a portability that the old-timers could only have dreamed of, and I am referring to many more types of telescope than just the popular (and predictable) catadioptrics. Our telescopic ancestors (only a wink in actual time before the present) had to fabricate most of their equipment themselves, and used to build their scopes from truly massive materials in order to obtain the stability they needed. When they wanted to go on location, they would lug enormous loads of heavy gear up to mountain tops in order to view the stars. Hard as their lot was, I bet nevertheless that they had more fun than we have today! Their dedication is still to be admired; without them, we would not have amateur astronomy in any capacity today. To me, their role proves that part of the fun is always the challenge, and therein is the key to a large part of the downside I see.

On the upside, modern computerization not only has assisted in the commercial manufacture of all of this equipment (and helped to lower costs), but also has been incorporated to a considerable degree within the functions of the astronomical hardware itself. Not long ago, just tracking something with reasonable accuracy was considered a feat in itself. We relied on drive correctors, exact polar alignments, and so on, even our manual skills, and were grateful if our mountings had slow motions on both axes, let alone that they might be electric! Digital circles have been a godsend to many, even to those who already know their way around the sky to a pretty good degree. I do not know what I would do without them these days. In their absence, to line up otherwise unwieldy telescopes on minute areas of the sky, such as in the case of my own scope, would require great patience. Certainly it would require more than I have most of the time! Schmidt-Cassegrain telescopes, and the like, can be especially difficult to use when sighting objects along their short, stubby tubes. Fast forward to the present and add perfect object location, "go-to" capability, and it is easy to see why this feature has become so popular in modern catadioptrics. Regardless of the type of telescope being used, the old mechanical setting circles do not even come close!

Some thoughts on Schmidt-Cassegrain (SCT) telescopes; in many ways remarkable tools for the amateur, they may not be all that their promoters claim, or at least, imply. The sophistication required for their manufacture has to be admired, something made all the more remarkable by their great availability. We have to consider, though, that the chief advantage they offer us is their well-known compactness and portability, particularly in the less than grand sizes. The SCT is certainly a compact design, far more so than those traditional F8 equatorial Newtonians of yesteryear. However, quite honestly, by the time we reach the 16 inch (41 cm) telescope by Meade, it is a considerably more massive instrument than most people realize. Setting it up is really a two-person affair; by contrast, I can set up my modern 18 inch (46 cm) JMI Newtonian alone. My JMI scope is also less bulky, and has a considerably smaller footprint than the large field tripod of the Meade.

We should look at SCT scopes, as a type, for a moment. Scopes of the highly practical sizes may be ideal for many observers, especially in view of their compactness compared to other telescope designs. They represent a reasonable solution to the challenges city-bound observers have when relocating to better sites, particularly to those people less able to haul and set up larger and heavier equipment. However, they do cost quite a lot more than the normally bulkier Newtonians of the same aperture, particularly so when they come equipped with elaborate on-board capabilities, including go-to object acquisition. This particular feature is anything but necessary, although with the inherent aiming problems with Schmidt-Cassegrains (from a mechanical standpoint), they may actually have some justification in practice. However, the manufacturers have apparently concluded that electronics should be the primary emphasis these days for this type of telescope. There is no denying that there should be extra appeal to these complete and highly portable bells-and-whistles telescopes for those whose interest is sufficient to spur some regular activity, but who maybe do not have the time needed to hone their skills and knowledge to a greater degree. No argument here; may they have fun. Maybe they will be so inspired they will graduate to ever-greater pursuits, astronomically speaking.

However, something needs to be said about the casual claims of optical excellence made about the SCT telescope configuration, as well as assumptions that the uninitiated might easily be led to conclude regarding performance. It is often stated that they are really good all-around instruments. This point would seem to be the first line of defense to deflect the age-old weakness of their design. While acceptable performers at most things, it is just as true that practically all other standard types of telescope do as well, if not better! I recently read a review of these telescopes, which rated them as "single instruments that perform well in all fields," as if most of us use a wide variety of telescopes for each aspect of our astronomical pursuits. While a particular and specific optical attribute or disadvantage with any type of telescope is actually quite hard to demonstrate (and I am referring to one that most uninitiated observers would actually notice through the eyepiece), I hope that "Jacks of all trades" is not the term that was actually meant in that reference to SCTs! Though I am not going to say they are "masters of none," you should know a little more about what you should expect to see through such an instrument. On solar system objects, they will indeed give decent if somewhat soft images, which may lack the contrast you seek. For deep space applications, they will do better, but contrast issues may again be critical at times on very delicate subjects. They do better with astrophotography, and best, more particularly, with CCD imaging, since contrast may be manipulated later to a great degree in computer image processing. In all cases, however, accurate attainment of focus also may be noticeably harder than with most designs. This is due to atmospheric disturbances interacting by diffraction with the large size of the secondary mirror, and the design's usual dependence on the position of the primary mirror moving, instead of the focuser itself. In addition, the tiny fine focus adjustments of their focuser units' design may prove tricky, compared to other types of telescope.

Now for the issue of secondary mirror size: when is the last time that you ever saw the dimensions of an SCT's secondary mirror volunteered in an advertisement, or even listed on a specifications sheet? The truth is that the central obstructions are the Achilles' heel of these telescopes, and perhaps no manufacturer would want willingly to advertise it. With the normal central obstruction of an SCT averaging over 35 percent, it was when I saw the casual concession in one recent review that an 8 inch SCT will only perform like an unobstructed 5 inch (127 mm) on solar system subjects that the truth about these systems was once again brought into the spotlight! This statistic is certainly not so good. Granted, Newtonians and others have central obstructions too, but not nearly so large and damaging as with the venerable SCT. In these other types, the obstruction is often scarcely an issue, particularly when it is well under 20 percent of the primary's aperture. Even a sizable 22 percent obstruction, such as found in my telescope, is only marginally detrimental, especially when good tube ventilation is built into the design. Most people would be hard pressed to tell the difference between this kind of system and a telescope with a vastly smaller secondary, or maybe even none at all. However, secondaries larger than this definitely create an issue, and by rapidly noticeable degrees as their dimensions grow.

Then comes the important aspect of compactness and portability. As the apertures increase, selecting a Schmidt-Cassegrain may actually pose more problems of bulk than makes logical sense to you, along with its less than ideal live viewing potential. It is with larger sizes that competition from more conventional optical designs begins to kick in, and in the best instances would seem to me to relegate the SCT design to a less valuable role. For moderate apertures, you may instead want to consider something like an advanced JMI equatorial Newtonian design, which although still more expensive than the most advanced models of SCT, will outperform them, and be easier to move to boot. Only when we reach the quite large (by amateur standards) JMI 18 inch Newtonian do we find that the reverse applies with cost, and the smaller Meade 16 inch SCT actually costs more. In my view, the JMI telescope is nevertheless a far superior instrument for live viewing, although it does not have the precise pinpointing and go-to capability of the Meade. However, this feature really is not needed by a moderately skilled observer with a telescope as comfortable to use as the JMI. We still have complete equatorial capability and digital circles, along with outstanding tracking with minimal periodic error, due to the large and accurate equatorial ring. So carefully considering what is right for you is very important, as you are gripped ever more by the incurable disease of aperture fever. I guarantee you will be infected with the bug, so you can save yourself considerable expense and disappointment by thinking ahead.

There is a fine line where wonder ceases and humdrum begins. Many of today's commercial scopes and equipment have made the universe seem no different from switching on the TV, particularly for the newly initiated. Much of the widely available astronomical equipment even looks something like a VCR, or some other consumer item! Different major manufacturers make telescopes, eyepieces, and mountings that are often hardly distinguishable. The Schmidt-Cassegrain and its derivatives reign supreme among the mass producers, as if no other type of telescope is worth considering. We finally have cookie-cutter astronomy! There is very little to find in this mass-produced market that seems to look toward superior performing optical designs, certainly in their fanciest and most touted telescopes; tube compactness and fancy electronic features still occupy the most sacred places. How many speeds of slewing can we find really useful, after all? Though the well-established SCT may fit the bill for many people's needs and enjoyment, the essence of astronomy seems to be taking a back seat to fancy features. There could be many potential enthusiasts who are getting shortchanged by experiencing something more technically wondrous but less inspiring in terms of what they will actually see. They will not stick around long enough to discover the other side of astronomy. The worst downside of all could be that live viewing will be gradually made to take a back seat to fully automated CCD imaging – almost like some type of interactive television. I guess I should not complain too much; at least it does not have commercials (yet).

Aside from those few daredevils in the amateur community who still strive to come up with something really unique in construction and design, which is otherwise utterly unavailable in the marketplace, one can hardly be blamed for thinking that there is no longer a whole lot of reason to become a telescope builder. With the rise in the standard of living, and the relative decline in the costs of astronomical gear, who can blame anyone for leaving alone the considerable challenge of building? Just go out and buy a pre-made scope, complete with presets and even preprogrammed sky tours. Not much to think about, just set it up and go. The wonderful creative and sometimes eccentric telescopic creations of Russell Porter's day seem to have almost disappeared, but also sadly with them, the dreams and passions of their creators.

However, you should know that it was certainly the creative part of astronomical pursuits that had such an early and permanent grip on me. This included fabricating telescopes with my own hands, making it such a personal part of finding my way to the stars. I am sure that this ingredient was also a significant factor in the astronomical craze that so swept all of those amateurs of earlier times. It often became a lifelong pursuit for them. (I will confess, however, that I found at times that there was a fine line separating the joys of building and the pleasures of actually using the crazy apparatus I had put together! But what memories and fuel it provided for the present!) Together with this entire process came the training, by default, in the actual use of telescopes, with an innate comfort level and familiarity with them. In addition, there was the invaluable experience in learning to extract the maximum there was to be seen in the views these creations provided. I think one is more likely to strive for this when it comes to using one's own fabrication; somehow, there seems to be a greater need to justify the hours spent building it!

Some dominant forces coming into amateur astronomy today are those relative newcomers to the field, whose primary link with us came through computers, and their own expertise in that particular field. These usually bright and energized people, very welcome in astronomy no matter what may be implied here (they offer us a great many new links with expanding technology), often seem to carry with them nevertheless a certain mindset that puts much of astronomy on a more clinical footing, what we might call a "hardware/software/double-click/pure educational" mentality. Very little magic is left by the time all of this has run its course, and it is here that we have suffered some of our largest losses from the spirit of wonder. Add to this the easy dependence on the Internet, rather than actual firsthand experience, as well as a generation accustomed to instant gratification, and you have another recipe for astronomy being reduced to a very uninspiring and even controlled subject.

Worse still is an elitism that pervades many amateur astronomical circles. Sometimes, in an apparent contradiction, it seems to fly in the face of technological advancement. It may be as simple as the outright condemnation of the use of digital circles, where only "star hopping" is seen as something that true believers are able or willing to do! (At least these people are still visually oriented!) Another example would be the widespread total rejection of the image intensifier, often without those same folks who condemn them having even tried one. I have already commented on this. The list goes on, but I think you may get my point. Some people have a greater wish to keep outsiders and new ideas out of their universe than to see it expand ever larger in a Big Bang of their own making.

I think I see also other signs of elitism at times in some of the high profile astronomical journals for the amateur enthusiast. Now, do not get me wrong; these magazines have done a heroic job in promoting all that is astronomical. I also understand the problem of providing new and varied information to readers on a monthly basis. But it has to be said that on occasion the mainstay articles of some of these periodicals center around highly esoteric and complex issues, as if the writers have forgotten who most of their readers are: it is the amateur observer, most of whom enjoy nothing more than simple hands-on astronomy, and not the professional or doctoral candidate. One only has to look at the multitudes of advertisements by every manufacturer of astronomical equipment to know this.

There is no doubt that cutting-edge issues are very informative to readers interested in pursuing a deeper grasp of the ongoing discoveries in astronomy, as well as the true student. It is also probably true that readers with even the most elementary level of interest enjoy perusing the same information at times, but not as their staple diet. So I would bet when there is an overabundance of such articles, to the average amateur observer, the direct connection with these and their own observing pursuits must seem vague or unconnected. This would particularly be the case when we consider the type of equipment most amateurs have at their disposal, or which they usually read about within these magazines' pages. Most average enthusiasts use instruments quite limited in potential and light grasp, and normally there is not much guidance to a grander level of viewing. So we have a strange contradiction: the major thrust of many telescope reviews in these periodicals centers around instruments far too limited in potential to connect with much of the feature articles in question, and certainly not with most live viewing. (Not that individually many of us can directly connect with the major thrust of the most esoteric of these articles anyway, no matter what equipment we have at our disposal!)

In other types of disconnect between subject matter and observer, how many amateurs really want to see detailed features about such things as life aboard the space shuttle, for example, when they open one of these magazines? While closely aligned subjects such as these may well be of interest, would it not be more appropriate in another type of magazine? Most astronomy magazine readers really would probably rather see articles that help them expand their own potential as amateur observers, and would be better served by efforts to keep their own pursuits at the forefront of the periodicals' focus and attention.

Prominently placed product reviews and articles in the same magazines describe not only equipment available in the marketplace, but the latest in astronomical computer software, and often the unwritten but clear implication (even acceptance?) that practically all things lead to CCD imaging. It often seems the unique value and experience provided by visual astronomy is vastly unrecognized or not promoted these days. Many telescopes under review feature all kinds of new elaborate electronic features, usually eliciting an almost predictable euphoria among the reviewers; however, in truth, very little about these various novel innovations has much to do with the real art and technique of amateur observing. Just randomly perusing a number of these magazines from the recent past, the apertures of telescopes featured in the majority of the reviews seldom exceeded 4 inches (101 mm)! Let us face it, even double that size is not exactly big any more; however, you are lucky to see anything larger than this (8 inches/ 202 mm) receiving much attention in reviews these days. It seems that most of them center around either low-end smaller instruments, or those of similar aperture at the diametrically opposite end of the cost spectrum. In many cases these high-end small, or medium small, telescopes average several thousand dollars and more. So the reason for favoring limited apertures cannot be due to considerations of cost! To me, the ultimate in what nevertheless have to be considered small telescopes, such as the majority of top-of-the-line apochromatics, as well as the latest in automated features and gadgetry, would seem far removed from the core of what made amateur astronomy what it is in the first place.

Actually, much of the commercially available, grander-scale equipment costs significantly less than the small high-end instruments often featured in these reviews. The alternative approach (of actually going for light grasp with excellent quality, over the much more limited capability of the very finest quality, small instruments) offers multiple times the live viewing potential, and actually has a real chance at connecting with perhaps just a little of those esoteric themes in the magazine pages. I will concede, though, that occasionally an article or even a review does appear in one of them that features some grand and unique telescope construction. This sometimes extends to a fine observing or related program by an amateur enthusiast. Compared to what so often seems the norm, it feels as if we have briefly reached astronomical Nirvana! I think we just need to reach it more often. This would be so much more welcome than the more usual fare of predictable or relatively mundane, even sometimes obscure, building projects. It would be so much more compelling than reading about instruments of limited concept, which often just rehash the basics of what we already know, albeit in a slightly different way. This is also true of the standard articles about CCD imaging, computer software (forgive me, in my view, some of the least interesting features ever to appear in astronomy magazines!), or what seem to be the almost obligatory and truncated observing pages, often looking as if they have been grudgingly included after everything else (of "importance"?) has been covered first! It does usually seem that actual observing and telescope-building features in these magazines are relegated to the back pages.

Since there is no shortage of quite affordable larger-scale equipment around today that really performs, could we not aim more frequently for greater practical performance values with what is reviewed, promoted, or featured? Also, how about many more high-profile articles in these magazines that feature the results obtainable with some of the grander scopes available or even those built by amateurs? I would love to read many more discussions on their actual use by more of those "out here." Therefore, this is a plea for more of our observing soul mates to be found and heard from in these pages, the vast universe of astronomically inclined people, otherwise unheard of or unseen. Indeed, here, CCD imaging could have a great connection with the live event, serving as a direct source of comparison with the live view, as well as showing better what to look for and extract from live observing. More particularly, articles featuring real-time imaging, such as I have engaged in or discussed for the purposes of this book, would also be an incredible resource for the live observer. Featured within regular and significant observing articles, this could provide new life and impetus to the fading field of visual astronomy, which right now seems on the fast track to oblivion, at least as far as some of the implications appear.

It does not always follow the pattern. A recent magazine article I saw in one of the most respected astronomical periodicals (unfortunately, an article relegated to the back pages!), did, however, bring great attention to drawing deep space objects from live viewing. It went to great lengths to contrast the differences between the live view and the time-exposed image. It was quite wonderful to read. Another recent prominent magazine featured as its mainstay, throughout, many articles on Mars at the time of its recent most favorable opposition, with emphasis on practical observations and recording, even drawing! How exciting and refreshing!

In an unlikely corner of the publishing galaxy, one small magazine in particular strikes me as possessing strongly the spirit of all that seems, to me at least, to be in partial eclipse. It features exactly the kind of articles that I find so frequently in short supply in many other places, along with attention to amateur observers' grass roots connections, and telescope building and the like. Going by the name *Amateur Astronomy* (www.amateurastronomy.com), this is the little magazine that could. Firmly keeping connection with the fundamental core of what drives the spirit of all things astronomical for amateurs, it also shows just what can be done outside the more common commercial prescriptions. Indeed, it contains something for everybody who is passionate and serious about hands-on astronomy, who is aware of the increasing emptiness of soul to which we find ourselves often being so routinely guided. It certainly does not fall into the orbits such as I have discussed, and is a notable exception to the commercial direction we see so often these days in every aspect of the hobby.

I should add that building a sizable or moderate-sized scope, even a home-built economy scope, is not a prerequisite to a lifelong interest in astronomy. However, it can greatly help with the direct link it establishes with all things astronomical. In my own case it has worked to the degree that I am sure that I have a totally different level of connection with my present telescope (commercially built though not mass produced) than I could have otherwise enjoyed had I not built telescopes of different sizes myself. As a former telescope builder, maybe I did not always meet all my criteria in my own constructions, but it was great fun trying! However, because of the time spent with building, I now know, firsthand, what is really important in telescope design and features, and certainly what was important to me. When it came to buying a telescope, it was easy to see what I needed in a telescope not of my own making, and exactly what I did not need in wasted features.

The fact remains, however, that many people simply are not adept in anything vaguely mechanical, and there is nothing wrong with this. The fabricating of scientific tools is not for everyone, no matter how simple. But I still want to stress that every person who gets involved with astronomy should get as much hands-on experience with as many types of scopes as they can, in order to acquire that kind of "sixth sense" that all of us who have worked with them over lifetimes pick up along the way. You will not acquire that sense by buying your ultimate telescope right off the shop floor. To do this would be akin to the person who has only driven one car, the ultimate, but has never actually driven any other, or perhaps never has worked on any type of vehicle; it would be hard to have any proper perspective. For those who do not have the kind of telescopic background I refer to, however it is acquired, we can only add to this that the limited feel many newcomers have these days is presumably the result. This effect is only enhanced greatly by the extreme automation and features, now becoming so common on the main commercial lines, and which are actually way beyond practical value for most. While these features are wonderful in many ways, in the wrong (or shall we say, uninitiated) hands they may have a considerable downside. Is one left to conclude that this is the only way that such telescopes can be marketed in the computer age, or are they designed just to appeal to the uninitiated, or those who have never spent any time at the eyepiece?

My own telescope could be described as about as unwieldy as any I can think of, yet with the use of its digital circles alone, finding any object is about as difficult and time-consuming as the finest and fastest fully automatic scope. Is there really any need for pointing accuracies calibrated to arc seconds in the most commercial and automated designs, or are the manufacturers perhaps telling us that CCD imaging is really where their market is, and that visual use is secondary, or not even an issue any more? Maybe some new users do not even know how the objects should appear, live in the field of view; perhaps they otherwise might not know what to center! Is this an unlikely scenario? Actually, I do not think so. Couple this with the automatic star alignment and GPS features becoming prevalent today: the user does not even have to know the name and location of any star in the sky at all, or actually to know anything! One can only assume that, in this regard, these instruments are being sold with the least likely customers in mind, those who will quickly discard their new toys once they realize that there is considerably more to seeing or imaging the wonders of the universe than may have been promoted. No matter; the sale has been made. Most experienced users would not pay premiums for features they really do not need; they just do not know what they need. Most of the features they pay dearly for will probably not fit their observing profile.

I know someone to whom all of the above applies. He has absolutely no idea what draws me to astronomy, and having been disappointed in his own experience, seems now unwilling to let me show him. Having bought an expensive, highly generic telescope (you can imagine the type), he became fired up with eager thoughts of doing some of his own stargazing. He had bought into the myth that astronomy was as simple as flipping the on-switch and seeing the universe readily displayed before his eyes. Of course he bought a glossy, general-purpose book on astronomy (one!), and he figured he had the complete kit. Additionally, because he did not even come to terms with the most elementary of instruction manuals that is supplied with the scope, his fancy new gadget still sits in his study, unused to this day. It must seem on a par with setting the controls of his VCR, which I am also sure is still flashing at him, yet to be programmed to this day! Because he has no background or inspired quest for astronomy, he has none of the fire to propel him to find the answers. He also seems to have no appreciation of what I have shown him through my own equipment, except that some things look nice. They do not look like the pictures he has seen. After all, there should be more to amateur astronomy than something akin to the purchase of just more consumer electronics. Yet that is what a lot of it has come down to. And how much value can there be to on-board catalogs of tens of thousands of objects, way too faint for these (usually too small) telescopes, even in the best dark sky conditions, to reveal as anything more than the minutest, faintest blip or smudge, if that?

Finally, forgive my mention of the multitude of commercially oriented, overproduced, color-saturated and repetitive astronomy books, which often lack the specifics we may be looking for (actually, much like the book purchased by the fellow I mention above), and which clutter your local bookstore shelves. Many of these same publications seem to ration really significant astronomical images of the subjects they cover (from the countless available, especially those imaged by spacecraft), in favor of splashy, overcolored graphics, empty visuals, and generic discussion. There really is a shortage of books that provoke the kind of intensive reading, thinking, and imagery as did some great epics of yesteryear, such as *Amateur Telescope Making*. (There are still some notable exceptions from publishers who do indeed understand the amateur's dilemma, and are continually trying to address the problem. Usually, one has to seek them out. Among them, I must recognize my own publisher, Springer Books, for its ongoing efforts to address the needs of the true amateur observer.) For those who have never experienced such inspiration, it may never be possible to catch it in the present climate, and so the lifelong inspired pursuit may be lost to something much more mundane and clinical. I hope these newcomers may instead have the opportunity to be influenced by some enthusiasts of the old school. I acknowledge that things do indeed change, but it is vitally important to keep one's inspired reason engaged, as well as that of creative application.

So there it is, my mixed narrative on the state of affairs existing in the amateur astronomical community today. I hope the downsides I see are not too depressing, but it is at least as I see it. For all that we have gained, you may agree with me that many new astronomical recruits (amazingly, even some old hands too!) are in danger of losing the big picture at the same time. This need not happen. The other side to it all still exists, but is just getting buried in the hype. Amateur astronomy always needs to be based on, and include, an enlightened, visual foundation. This is never clearer than when you see the handiwork of many of the supposedly advanced CCD enthusiasts. The grainy gaudiness of their planetary images illustrates most graphically how little time they have ever apparently spent at the eyepiece of a telescope!

It is not too late to save the golden age of astronomy for everyone, never too late to show what can be done. Observers such as me will always be inclined to look through the eyepiece first. If we of like mind do our part in pointing others toward the out-of-this-world legacy entrusted to our eyes, it will not only survive but flourish. For us, if this means that we will always need to adapt and evolve our approach to take advantage of new visual technology, this presents us with nothing but pluses. It can only expand our potential. For some, alternative forms of live viewing, or advanced imaging, may of necessity be substituted for the eyepiece itself in the final end game. However, no CCD enthusiast should allow himself or herself to be raised as anything but a visual observer first, even if actual real-time visual observing is ultimately put on the back burner for years at a time! One such amateur observer I know, Jerry Keegan, Special Projects Marketing Manager of Scope City stores chain in the western United States (where I purchased my NGT-18), still pursues both fields with a passion, never confusing the special values of each. He actually boasts an eyepiece collection worth more than $8,000 (£4,348 UK), despite the breadth and sophistication of his CCD equipment!

So, for me, the quest remains the same: to preserve the spirit of amateur astronomy. As long as visual observers are around, they will not let their part fade from view. But they must be around! They must not allow themselves to be talked out of its future by others, no matter how learned these folks may appear to be. So, having by now made a lot of people unhappy, let me try to redeem myself(!), and offer a few thoughts about telescopes that may best suit your specific needs. You may be surprised to see that some scopes may even be among those I have apparently condemned. As the ultimate deciding factor, we will need practicality, and especially personal suitability, above and after all. You should know that it is all right if some of your specific needs violate some of those apparent tenets that I have just propounded so vociferously! Anything we choose must fulfill certain requirements for us personally, regardless of my previous comments; so, there are no rules. However, there is nothing like informed decision. And while I still do not intend this book to be yet another review of the multitudes of instruments on the marketplace, I think it is appropriate to examine some important considerations, including a brief look at certain commercial products well suited to our task and individual preferences. Some rare enthusiasts will wish to build; more power and enjoyment to them. Many more, though, will decide to buy.

While you may be capable of fashioning a fine telescope of your own design, it may nevertheless be built in such a way that makes it extremely difficult or impossible to relocate to a remote site. If you wish to follow in the great tradition created by those inspired folks of yesteryear, it is a wonderful quest, but you will need to pay special attention to portability, in ways those pioneers never had to consider. Years ago, there were countless reasonably dark sky sites, even within communities. It was not so hard to install a scope on a hilltop somewhere nearby or in a dark backyard, provide it with functional shelter, and unveil it whenever the sky was good. As time has marched on, the ever-encroaching expansion of Edison's mighty invention (the lightbulb!) has slowly but very surely robbed us of most of our pristine black skies. Therefore, unless you live where the skies still get truly dark, and you wish to see the best that the sky has to offer, you are going to need to move your equipment to remote locations more than once in a while. This has to be a practical activity, otherwise you will do it only once. Some observers have considerable means of transportation at their disposal, such as trailers, trucks, and the like, but this may not be in the cards for everyone, or even the best approach. Without such means, whenever you wish to relocate to faraway sites, this will entail looking for all possible ways to reduce weight and bulk. However, it should not be at the expense of stability, or a telescope approaching your ideal telescope aperture. If you want to have tracking capability, this only complicates things still further, but there are some excellent examples of solutions to these issues that have been undertaken by amateur builders. But I think it is fair to say that the era of massive and permanently anchored "truck axle type" telescopes has come and gone for most people.

It is also a sad fact of life that some of the great commercially built telescopes of yesteryear may also prove to be of limited value these days for many observers, considering their bulkiness and sometimes massive weight. The problems of relocation to the dark skies, which many of us face today, may make them a difficult option. I am referring, of course, to now-legendary manufacturers, such as Cave Optical Company, the Optical Craftsmen, Coast Telescopes, even the lesser-known Charles Frank of Glasgow, Scotland, or any of the great refractor makes of antiquity. For example, the equatorially mounted Frank 6 inch (152 mm) would rate, in my opinion, as probably the single finest telescope of its size ever made. There were, of course, many other makers of high-quality telescopes in an age where commercialism had barely taken on astronomy. You would always do well to consider any of these wonderful old telescopes, but only if it can fit the parameters of your own viewing conditions. Unfortunately, for many observers, the size and bulk of any of these mostly German equatorially mounted (usually F 8 reflectors, or typically F 15 refractors) would preclude easily relocating the larger sizes to dark skies from their city confines. What a pity; there has scarcely been anything of their workmanlike, solid design and construction ever made since. However, those lucky amateurs who already live under dark skies may have the best of all worlds, if they can obtain one of these splendid instruments on the secondhand market. It is not so difficult to upgrade any of them to take advantage of some of the new innovations.

I finally made the concession that I would buy my present telescope, rather than build another, after a life of telescope building. I cannot tell you that it was an easy decision, but in my own case, the JMI NGT-18 I purchased truly answers the needs I have today. Every issue has been practically addressed in ways that amateur builders would strive for, but done with the resources of proper engineering and design. There is no overriding commercial agenda in the execution of this telescope, except just making it as well as it can be done and still make a profit; remarkable indeed! Other outstanding examples exist in the marketplace, and we will get to them, but let us not rule out the commonest option available. By this I am referring to the highly commercially successful Schmidt-Cassegrain design, which has dominated the market for many years. You are already aware that there are some considerable pros and cons to these telescopes. You may end up electing this design as yours; countless observers have. While there are some compelling reasons to do so, I think you would be wise to look at the whole picture to confirm if this is really the best way for you to go. A word of warning: I have firsthand experience of one prominent and near-legendary manufacturer of such telescopes, now pulling itself out of potential collapse. While their equipment may indeed be among the best of the mass-produced commercial lines, when a warranty problem arose their customer and technical service was virtually nonexistent. They never resolved the problem and I was left to find other solutions. Needless to say I would not buy from them again, nor recommend them. You may want to research any company whose products are under your consideration.

Still think you cannot afford a large-aperture telescope? A visionary named John Dobson showed that huge apertures and startling results could be obtained really cheaply by those willing and open-minded enough to give it a go. You maybe know what I mean: Dobsonian telescopes. These telescopes dared to thumb their noses at the establishment, concerning what was feasible or acceptable. All of those things that could not possibly work, or so it was thought, were put to use. Made from all sorts of unlikely components, ranging from porthole glass, two-by-fours, even to cardboard and discarded junk, these offenders have two essential ingredients: stable design concept and simplicity of execution. Most are built with at least some degree of portability in mind, especially with their low-slung centers of gravity, allowing a compact mounting close to the ground. They also feature large turning faces on both of their axes. However, the larger telescopes usually need trucks or vans; Dobsonians were not conceived with compactness as their primary goal. John Dobson was apparently embarrassed that such a primitive apparatus had been labeled with his name, to be known as a separate telescope species, but he should not be. Their effect has been revolutionary.

Commercial manufacturers ultimately followed the lead with simple Dobsonian designs of their own, and there are some excellent examples available at low prices today; these prices will give any other scope a run for the money in pure performance dollars. However, I do not see more than occasional attention being given to Dobson's approach in any of the monthly periodicals these days (grand and simple construction, stability of mechanical design), and going for the maximum with the minimum investment. The truth is that his approach and passion for live observing would blow most of today's empty commercialism away, and fire up new amateur enthusiasm. The actual altazimuth concept of Dobson's design need not even be adhered to if something equatorial in design can be constructed equally simply. However, there is another way to attain equatorial capability. Let us look at the whole scenario.

The type of sophisticated equatorially mounted telescope that JMI manufactures still has something in common with simple Dobsonians of comparable or even larger apertures: they are all low-center-of-gravity Newtonians. In other available low-cost Newtonians, or even the more expensive Dobsonian designs, ready equatorial tracking capabilities do not exist, but for value and optical performance, they still may be the best bet for most observational astronomers. (By "more expensive Dobsonian," I mean one that takes the principal design features of the original concept, but refines the final telescope into a relatively sophisticated instrument. They frequently feature skeletal truss tubes, as well as a design that allows a breakdown into relatively small packages, with good thermal properties, keeping the weight to a minimum.) These telescopes can be equipped with digital circles, and fortunately, various manufacturers now offer equatorial tables (Poncet Mountings), which, although not the same thing as a true equatorial mounting, do allow real equatorial tracking for moderate periods of time. Probably the best of these equatorial platforms would be those made by the company Johnsonian Designs, very stylish and well finished, to say nothing of the quality of their manufacturing. Other types are readily available, or may be constructed for oneself. Comprehensive instructions and information may be found on the Web at http://www.atmsite.org/cortrib/shaw/platform/. This should answer almost any question.

Such equatorial platforms may be made by the amateur builder as well, the entire Dobsonian telescope and mounting – even an observatory – riding on top of them, tracking the skies by replicating a short segment of the polar arc that adjusts to right ascension. They work by allowing full movement and object location by the telescope's own mounting, but subjecting the whole assembly to a motion aligned with the polar axis. Tracking periods of an hour or more may be successfully undertaken, even allowing for photographic or CCD time exposures! However, digital circles will not work with the equatorial platforms.

So for value in performance, without actual any compromise of the really important features, the Dobsonian is therefore hard to beat, although I must concede I would be reluctant to trade the full equatorial capability that my JMI scope provides for one of them. I suppose that makes me spoiled, but JMI made me so! Companies that have established some of the best reputations in advanced Dobsonians would include Obsession, still at the top of the heap. This company offers some truly huge telescopes of a simple but utterly perfected design and quality. Then there are StarMaster Telescopes, Starsplitter Telescopes, and Discovery Telescopes, all of which make excellent truss designs, and while maybe not quite the equal of Obsession, are quite a bit cheaper and still of a magnificently high order of construction and performance. There are others, but beware of designs that have their own limitations, in particular those that feature large ball-like tube bases in which the main mirror rides. This ball-like base of the tube also serves as the major pivot point in the design, and while offering great ease of rotation with good stability, telescopes of this variety cannot be fitted with digital circles, a major shortcoming in my view.

With any open tube telescope you will need to invest in a light shroud, a simple black cloth curtain that makes your truss design become effectively a solid design, except without the disadvantages of weight, tube currents, and the greater cool-down requirements that solid tubes usually need. Other pluses of these shrouds are that they provide light baffling (obviously less important at dark sky sites), but they do help to keep dew off the optics, dust out (really important in the desert!), and provide some deterrent for the observer's own warm air currents from crossing the optical path. Unless you have no problems with portability and transportation, steer away from solid tube designs anyway; they will often be simply too long and bulky without some means to break them down. German equatorial mounts may also pose problems with the larger sizes for the same reason, as well as their much higher centers of gravity, but you must decide which is best for your own needs. Whatever you do, just make sure that lack of portability is not the reason your telescope sits at home, never to be transported to dark sky country. We are trying to encourage ourselves to undertake these adventures, not to give us better ammunition to make excuses why we should not!

Special mention should be made of a type of advanced Dobsonian gaining prominence, not only for its uncompromising performance, but its unique design. This is the beautifully crafted and truly portable Teleport line of telescopes. Originated in Texas, the manufacture of these telescopes has been taken over under license by Bray Imaging Technologies Ltd. of the U.K. This company offers (unfortunately no larger than these) 7 inch (178 mm), 10 inch (254 mm), and 14.5 inch (368 mm) models, all of a telescoping truss design that collapses into a self-contained package that is easily transportable. The collapsed unit is like a large cube, and even has an incorporated shroud that expands up the trusses as the telescope is expanded to operating dimensions. The downside is that they are expensive (the 10 inch being even more so than JMI's 12.5 inch equatorial), but nothing good comes without cost, particularly at the less than mass-produced end of the spectrum, such as these happen to be. Perhaps the greatest selling point is the really easy setup and breakdown, not to mention ease of transporting; this matters more than you realize after a long night out in the cold.

The most important thing to take away from this discussion is to consider more than just the most commercially promoted options for your own needs, or commonly accepted notions of what amateur astronomy has to be or where it is going. A certain part of the responsibility for where we are going must be taken and led by the periodicals. Many newcomers look to these sources for direction. Those of us with long roots in amateur astronomy know there is more to it than just the most commercially promoted options; there is an entire universe of awareness than could be presented for us to ponder and become inspired by, and to connect with more people like you and me.

In conclusion, I wish you many years of sleepless nights as you follow your dreams under the stars. Keep them alive. May you share them with many others.

APPENDIX 1

Recommended Equipment Manufacturers for Amateur Astronomers

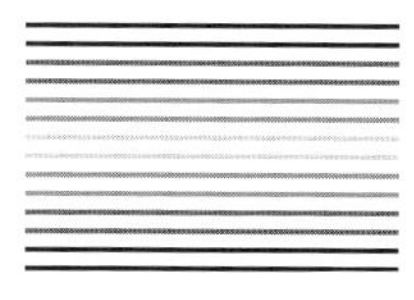

This resource is included here as a guide to those interested in pursuing image enhanced astronomy. This Appendix reproduces and adds to much of the information that was also contained in *Real Time Astronomy in the Suburbs*. Since so much of it is specific to the focus of this writing, it should be instructive to make those relevant parts of it available here as well.

By listing only certain manufacturers, I do not mean to imply that other outstanding manufacturers do not exist. Whenever possible, I have listed some international distributors of American products, or other international manufacturers. In other instances, I have indicated whether a particular company handles exports, if they do not have an overseas distributor.

Prices

I have listed approximate prices in the United States. I have also listed prices in pounds sterling. Some items are made or distributed worldwide and are available therefore without shipping costs and import duties. Generally speaking, it will be clear to which items shipping will apply. In these cases, overseas readers should expect to add shipping, the costs of which vary wildly according to carrier, speed of delivery, and destination, as well as adding import taxes. In the UK, a simple method approximating the price in pounds sterling for imported goods would be to add approximately 30 percent to the U.S. price, for shipping and import taxes, then divide that number by 2, reading the result in pounds, as shown here. Similarly, one could evolve a ratio for any other currency and import conditions.

Manufacturers of Image Intensifiers for Astronomy (USA)

1. COLLINS ELECTRO OPTICS (Complete systems for astronomy, ready for astronomical use; no adaptation necessary)
   9025 East Kenyon Avenue
   Denver, CO 80237
   (303) 889-5910
   www.ceoptics.com

   Unfortunately, Collins has no distributors overseas. Presently, certain export restrictions apply; contact Collins Electro Optics for export information for your situation.

   Intensifier products by these manufacturers will require adaptation for use in astronomy; some supply only intensifier tubes:

2. ELECTROPHYSICS CORPORATION (Generation II, III, IV units)
   373 Route 46 West
   Fairfield, New Jersey 07004-2442
   (973) 882-0211
   www.electrophysics.com

   Electrophysics exports to many countries; some European distributors include:

   AM Vision
   The Old Schoolhouse
   Wilberfoss, York
   England YO41 5NA
   (0044) 0 1759 388235

   Jabsco
   Ostsrabe 2B
   D022844 Norderstedt
   Germany
   (040) 53533730

   Jenoptec
   12, rue J-B Huet
   Les Metz
   78350 Jouy en Josas
   France
   (33) 01 34659102

3. D & VP CORPORATION
   P.O. Box 54074 N. Salt Lake
   Utah 84054-0274
   (801) 299-8548
   www.dandvp.com
   or: www.nightvisionweb.com

   D & VP has no European distributors, but exports most of their products.

4. STANO COMPONENTS, INC.
   P.O. Box STANO
   Silver City, Nevada 89428
   (775) 246-5281/5283
   www.stano.night-vision.com

Stano does not export to countries outside the United States.

5. ASPECT TECHNOLOGY AND EQUIPMENT, INC.
   811 East Plano Parkway, Suite 110
   Plano, Texas 75074
   (800) 749-3802/(972) 423-6008/7717
   www.aspecttechnology.com

   Contact Aspect directly for possible exports and dealers in Europe.

Sampling of European Companies Supplying High-Quality Image Intensifiers

1. OPTEX (Gen III equivalent miniature systems, probably ideal for astronomy)
   20-26 Victoria Road
   East Barnet
   Hertfordshire
   England EN4 9PF
   contact: simon@optexint.com
2. THE HOUSE OF OPTICS (Russian intensifier units incl. Generation III)
   Hunstanton
   Norfolk
   England
   07879-214651
   www.houseofoptics.ltd.uk
3. DELFT INSTRUMENTS NV
   Röntgenweg 1, 2624 BD
   P.O. Box 103, 2600 AC Delft
   The Netherlands
   +31-15-2-601-200
   www.delftinstruments.com
4. EURECA
   Messtechnik Gmbh
   Am Feldgarten 3
   50769 Köln
   Germany
   www.eureca.de
5. PROXITRONIC
   Robert-Bosch-Strasse 34
   D64625 Bensheim
   Germany
   www.proxitronic.de

For a *worldwide search* of other image intensifier manufacturers (or anything else!), log onto www.kellysearch.com, and type “image intensifiers” into the field on the home page.

Image Intensifier Suppliers Known Also for Recursive Frame Averaging Equipment

1. COLLINS ELECTRO OPTICS
   (see under Image Intensifiers-USA)

2. POYNTING PRODUCTS, INC.
   P.O. Box 1227
   Oak Park, IL 60304
   (708) 544-9188
   www.poynting.com

3. MEDELEX (incl. recursive frame averagers)
   732 N. Pastoria Ave.
   Sunnyvale, CA 94085
   (800) 644-0692
   www.medelex.com

4. MICRO SYSTEMS, INC.
   1502-4 Kwangyang 2-Dong
   Dongan-Gu, Anyang-Si
   Kyunggi-Do
   S. Korea 431-062
   82-34-421-0314/5
   www.microsystems.co.kr

   Contact these companies directly for export inquiries.

Typical Approximate Costs in United States (US Dollars) for New Equipment

Fully dedicated Collins Generation III system: $2300/£1495

Generation III intensifier tubes: $1000–$2000/£650–£1300

Generation II (intensifier only – will require eye lens & adapters): $500–$850/£325–£552; more advanced Generation II products up to $2000/£1300

Typical Approximate Costs in United States (US Dollars) for New Intensifier Accessories

CCD video camera and adapter for intensifier (Astrovid 2000 or similar/Collins I_3): $1000/£650

Media converters and software for interfacing with computers (Sony; also available from Adirondack): $400/£260

Infrared bandpass filters (for light-polluted areas): $250/£162

Recursive frame averager (Collins): under $1,000/£650

Special 50 mm F1.3 (1X) primary lens (Collins): $295/£192

Other lenses available include 12 mm, 25.; also lower focal ratio (F.95) for superwide-field use: $200–$800/£130–£520

CCD Video Cameras for Astronomy, and Accessories

1. ADIRONDACK VIDEO ASTRONOMY (also sells Collins image intensifiers)
   26 Graves St.
   Glens Falls, NY 12801
   (518) 812-0025
   www.astrovid.com

   Adirondack's distributor in the UK is:

   True Technology Ltd.
   c/o Nick Hudson
   Woodpecker Cottage
   Red Lane
   Aldemaston, Berks
   England RG7 4PA
   (44) 01189-700-777
   www.trutek-uk.com
2. SANTA BARBARA INSTRUMENT GROUP
   147-A Castilian Drive
   Santa Barbara, CA 93117
   (805) 571-7244
   www.sbig.com
3. INTERNET TELESCOPE EXCHANGE (also custom Maksutov-Newtonians: apertures to 16")
   3555 Singing Pines Road
   Darby, Massachusetts 59829
   (406) 821-1980
   www.burnettweb.com/ite

   ITE ships worldwide; its products are also available in England through:

   SCS Astro
   The Astronomy Shop
   1 Tone Hill
   Wellington, Somerset
   England TA21 OAU
   (44) 1823-665510
   www.scsastro.co.uk

4. SONY (Video Walkman; media converters)
   The Web site will direct and connect you to the Sony Web site of your country, and supply names of dealers in your local area, worldwide: www.Sony.net.

Typical Costs (in US Dollars) for New Video Equipment

CCD video cameras (Astrovid 2000 or similar, StellaCam EX or II): \$595–\$795/£386–£517

Complete SBIG-STV system: \$1995/£1297

Telescope and Accessories Manufacturers Mentioned in Text

JMI (superior split ring equatorial Newtonian truss telescopes; accessories for all makes, including motorized focus units)
810 Quail St., Unit E
Lakewood, Colorado 80215
(303) 233-5353
www.jimsmobile.com

In Europe, JMI products are available through:

Broadhurst, Clarkson & Fuller
6, Tunbridge Wells Trade Park
Lonfield Road
Tunbridge Wells, Kent
England TN2 3QF
(44) 2074-052156
www.telescopehouse.co.uk

Intercon Spacetec
Gablinger Weg 9
D-86154 Augsburg 1
Germany
(49) 8214-14081
www.intercon-spacetec.com

La Maison de l'Astronomie
Devaux-Chevet
33-35, rue de Rivoli
Paris, France 75004
(31)1427-79955
www.maison-astronomie.com

OBSESSION TELESCOPES (Superior Dobsonians)
P.O. Box 804s
Lake Mills, Wisconsin 53551
(920) 648-2328
www.obsessiontelescopes.com

Obsession sells and exports direct worldwide, but has no distributors.

ASTRO-PHYSICS (Superior Apochromatic Refractors)
11250 Forest Hills Rd.
Rockford, Illinois 61115
(815) 282-1513
www.astro-physics.com

Astro-Physics exports direct to Europe and many other countries, but does not export to countries where there is a distributor. Distributors include:

Baader Planetarium KG
Thomas Baader
zur Sternwarte
82291 Mammendorf
Germany
(081) 458802

Medas S.A.
57, Avenue P. Doumer
B.P. 2658
03206 Vichy Cedex
France
(04) 70-30-19-30
www.medas.fr

Unitron Italia Srl.
Giovanni Quarra
via Agostino Lapini, 1
50136 Firenze
Italy
(055) 667065
http://www.untronitalia.it

ORION (Maksutov-Newtonians, achromatic and apochromatic refractors, and low-cost, high-quality Dobsonian Telescopes up to 10" aperture, eyepieces, filters and accessories)
P.O. Box 1815-S
Santa Cruz, California 95061
(800) 676-1343
www.telescope.com

Orion does not directly export its products internationally, but does have two distributors in England:

SCS Astro
The Astronomy Shop
1 Tone Hill
Wellington, Somerset
England TA21 OAU
(44) 1823-665510
www.scsastro.co.uk

Broadhurst, Clarkson & Fuller (see JMI)

STARMASTER TELESCOPES (Dobsonians)
Rt. 1, Box 780
Arcadia, KS 66711
U.S.A.
(316) 638-4743
http://www.icstars.com/starmaster/

Although Starmaster has no overseas distributors, it is now accepting orders internationally.

STARSPLITTER TELESCOPES
3228 Rikkard Drive
Thousand Oaks, CA 91360
U.S.A
(805) 492-0489
www.starsplitter.com

Starsplitter ships worldwide; has no overseas distributors.

DISCOVERY TELESCOPES, INC.
615 South Tremont Street
Oceanside, CA 92054
(760) 967-6598
www.discovery-telescopes.com

International distributors include:

The Binocular and Telescope Shop
55 York St.
Sydney, Australia
011 61 2 9262 1344
mike@bintel.com.au

Grab Astrotech
Wiesenstrasse 6
D-74821 Mosbach, Germany
0 6261 670015

TELEPORT TELESCOPES
c/o BRAY IMAGING TECHNOLOGIES LTD.
Cherrycourt Way
Leighton Buzzard
Bedfordshire, LU74UH
U.K.
+44 (0)1525 219100
www.brayimaging.co.uk

Contact directly for export information.

Dobsonian Equatorial Platforms

JOHNSONIAN DESIGNS
3985 S. Lincoln Ave
Loveland, CO 80537
(208) 863-5518
http://www.johnsonian.com

Contact directly for export information.

EQUATORIAL PLATFORMS
15736 McQuiston Lane
Grass Valley, CA 95945
(530) 274-9113
e-mail: tomosy@nccn.net

Contact directly for export information.

Manufacturers of equatorial platforms in the UK are unknown to me at this time. Comprehensive instruction for building one's own may be found on the web at: http://www.atmsite.org/contrib/shaw/platform/.

Typical Costs (in US Dollars) for New Equipment

Johnsonian: models for 6″–25″ (152 mm-635 mm) from \$495–\$1495/£321–£971

APPENDIX 2

Resources

Prominent Amateur Astronomical Associations and Links

ALPO (Association of Lunar and Planetary Observers)
www.lpl.arizona.edu/alpo.com

British Astronomical Association
www.ast.cam.ac.uk/~baa.com

Astronomical Society of the Pacific
www.astrosociety.org
(Web site supplies listing worldwide of astronomical organizations etc.)

The Belmont Society (amateur astronomers' site; space object listings)
www.belmontnc.4dw.net

Sky & Telescope (Sky Publishing/astronomy information) Web site:
www.skypub.com

International Supernovae Network
www.supernovae.net

American Association of Variable Star Observers
www.aavso.org

International Dark Sky Association
www.darksky.org

Dark Sky Directory
http://ourworld.compuserve.com/homepage/pharrington/Dssd.htm

GSI, Italy (dark sky sites in Italy)
http://www.gsi.it/astronomia/darksky/index.htm

U.S. Naval Observatory (http://aa.usno.navy.mil) daily almanac:
http://aa.usno.navy.mil/data/docs/RS_OneDay.html

Astronomical League
www.astroleague.org

International Meteor Organization
www.imo.net

International Occultations Timing Association
www.lunar-occultations.com

NASA Photographic/Information Reference Web site
www. images.jsc.nasa.gov
(A comprehensive catalog of NASA's photographic records of lunar, planetary, and deep space subjects, together with links to many other relevant sites, including amateur images, groups etc.)

APPENDIX 3

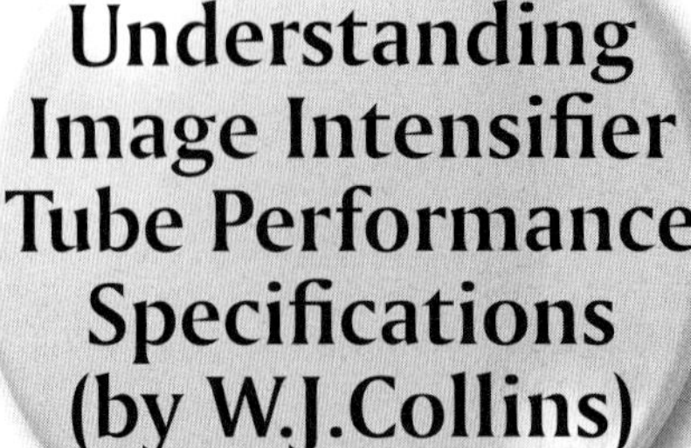

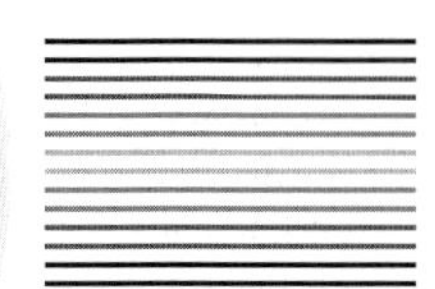

Understanding Image Intensifier Tube Performance Specifications (by W.J.Collins)

1. Signal-to-Noise Ratio (SNR)
   The most fundamental parameter for defining image intensifier tube performance, the SNR, is expressed in decibels (db), which are defined by the logarithmic relationship: db = 10log signal 1/signal 2. For example, if an image intensifier tube has an SNR that is twice as great as another, we define the intensifier with the higher SNR as signal 1; therefore (x) db = 10log 2/1 or 10log 2, which equals 3 db. The relationship of SNR to db is easily understood as follows:

   > For every increase in SNR of 3 db, the apparent visual noise in relation to the apparent brightness decreases by 50 percent.

   This concept is important. For the SNR to increase by 3 db, the image brightness in relation to the visual noise must increase by a factor of 2. For a 6 db increase the brightness must increase by 4, and so on. It is important to understand that "noise" refers to scintillation, also called shot or photon noise. The presence of noise is inescapable in a photo-multiplier system such as an image intensifier, and its primary sources are system amplifier noise and signal noise produced by the photons arriving at the photo cathode. Thermal noise, also termed Johnson noise, is also present in any detector whose temperature is above absolute zero and is manifested as localized system voltage gradients causing amplifier current fluctuations, adding to the appearance of visual noise. Since thermoelectric or Peltier cooling is rarely employed in image intensifier tubes, operation at cold temperatures will reduce the apparent visual noise. In simple terms, image intensifiers exhibit higher signal-to-noise ratios at zero degrees Celsius versus 30 degrees Celsius.

   During the recent past, "thin film" and "filmless" photo cathode technologies have given rise to dramatic increases in SNR as compared to previous intensifier tubes. SNR levels now approach 30 db, which is comparable to color video cameras in the not-too-distant past. Considering the enormous gain (as high as 60,000) of Generation III image intensifier tubes employing these technologies, such high SNR levels are truly remarkable. As important the SNR benchmark is, as we examine the additional parameters that define intensifier tube performance, we will find that tube performance is an amalgam of the combination of many specifications working in concert.

2. Resolution
   Intensifier tube resolution is measured in line pairs per millimeter (lp/mm), not to be confused with system resolution expressed in cycles per miliradian (cy/mr). Generation III Omni 4 intensifier tubes have 64 lp/mm resolution, which compares favorably with many "glass" optical systems. Such high resolution is due in large part to advances in microchannel plate technology, specifically pore size reduction to the 6 micron level found in Omni 4 specification tubes. High-definition Generation II intensifier tubes, which are available internationally, also incorporate 6 micron pore size microchannel plates, producing a resolution of 57 lp/mm. The Generation III Omni 4 and Generation II high definition (also called Gen 2+) have 18 mm photo cathodes and 18 mm phosphor screens. At 64 lp/mm we have 1152 line pairs resolvable across the diameter of the phosphor screen. Although image intensifiers are technically analogue devices, we can derive a digital analogy by comparing the resolution of an integrating CCD video camera with a " format CCD detector. The pixel count for such cameras is typically 768 × 494 pixels, totaling 379,392 pixels. For our comparison, we will use 1 line pair as the equivalent of one pixel. This is reasonable based on the Nyquist theorem that data (in this case 1 pixel) is sampled at twice the maximum frequency being observed (i.e., 1 line pair per pixel). Now, if we take the area of the phosphor screen in square millimeters as 254 square millimeters (area of 18 mm phosphor screen in square millimeters) we have the equivalent of a square 15.94 mm on a side. Therefore 15.94 × 64 lp/mm = 1020 lp per side. 1020 × 1020 = 1,040,400 line pairs over the total available area of the phosphor screen. Again using our line pair/pixel equivalency, we find 1,040,400/379,392 = 2.74 times the available picture elements on the intensifier versus the CCD video camera, a dramatic improvement.

3. Figure of Merit (FOM)
   This is a tube classification specification derived by multiplying the resolution in lp/mm times the SNR. The FOM is primarily used for determination of export requirements and is assigned a maximum (1600 in 2004).

4. Photo Response (PR)
   PR is specified in microamps per lumen. The gallium arsenide photo cathode used in Generation III image intensifiers has its peak sensitivity in the near-infrared band at approximately 800 to 820 nanometers. The curve of PR can be seen in Figure A.1, which plots wavelength in nanometers versus output in microamps per lumen. The photo response is correlated with the intensifiers quantum efficiency (QE), which is the ability of the photo cathode to convert incoming photons to electrons. The peak QE is found at or near the peak of the PR curve and may approach 50 percent.

5. Equivalent Background Illumination (EBI)
   The illumination level present at the phosphor screen with no photons present at the photo cathode. The EBI is affected by temperature with higher temperatures producing higher EBI. The EBI is measured in lumens per square centimeter lm/cm2). The EBI level determines the zero contrast level between low-brightness objects being observed and the inherent background illumination level. The EBI (measured at the photo cathode) with tube gain level in the 50,000 range is typically 3 to 6E-12 lm/cm2.

6. Halo Diameter
   The largest diameter measured in mm on the phosphor screen of a bright point source. A bright star is a perfect model for measuring halo diameter, which has a maximum specification in Generation III Omni 4 tubes of 1.5 mm.

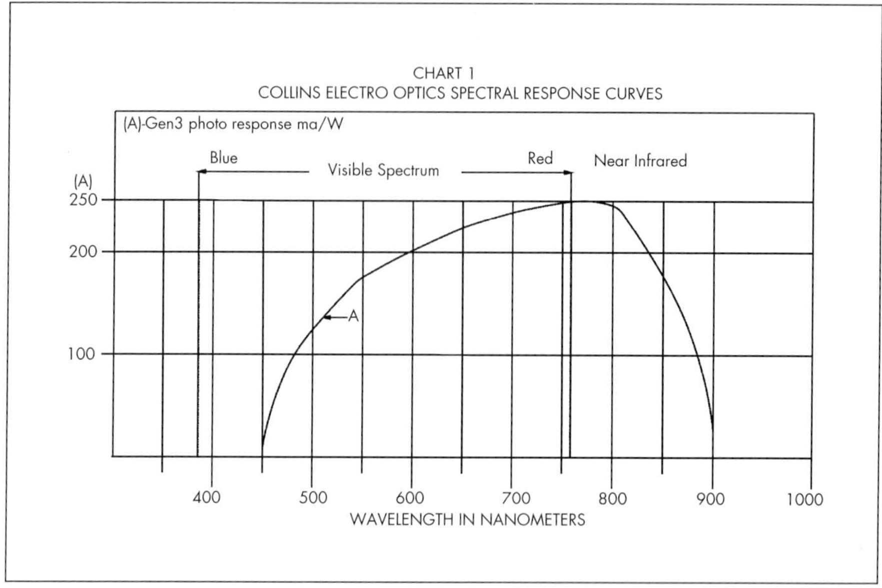

Figure A.1.

7. Modulation Transfer Function (MTF)
   Graphically expressed in a curve that plots image contrast in percent versus resolution in lp/mm, MTF is not generally found as a specific parameter for image intensifier tubes. MTF is, however, the defining optical performance criterion for any optical system in terms of overall performance.

8. Real Time
   Image intensifiers are true real-time devices, meaning an image created by the prime focus of any optical system with its prime focus coincident with the plane of the photo cathode will be observed with nearly zero time lag at the phosphor screen (the time lag can be measured in nanoseconds). Any other device that incorporates a CCD detector must accumulate photons over a period of seconds to reach the low illumination sensitivity of a Generation III intensifier, particularly a thin film state-of-the-art tube.

When ITT and Litton introduced filmless photo cathode technology several years ago, the "Generation IV" name was immediately associated with this new technology. It comes as a surprise to most people that the Department of Defense has never officially recognized the "Generation IV" designation, and in fact the term Generation IV as it applies to today's image intensifier tubes is of no relevance. All military products as shown on the ITT Night Vision Web site (www.ittnv.com) are described as Generation III. The differences between the military Gen. III tubes and the Gen. III tube in commercial Generation III image intensifiers such as the I Cubed (I_3) are as follows:

MILITARY TUBES

1. *Gated power supply:* The gated power supply in military tubes cycles the photo cathode between on and off states at a high frequency to dramatically reduce blooming in the image during weapons fire, for example. This has no relevance to astronomy unless you want to observe deep sky objects during a fireworks show.

2. *Filmless photo cathode:* Another measure to reduce blooming to a remarkably low level (almost nonexistent in these tubes). Same caveats as above.*bl/NL*bg

COMMERCIAL TUBES

1. *Nongated power supply:* This is the power supply in the I Cubed. The antiblooming requirement does not exist for astronomical applications.
2. *Thin film photo cathode:* Not for applications where bright source image transients may occur (like fireworks), although the maximum halo diameter is twice as good as previous MX10160 Generation III tubes.

The most important criterion for comparison is performance. The Generation III F9800 Select Grade thin film tube in the I Cubed has identical performance specifications in terms of signal-to-noise ratio, photo response, and resolution, as compared to its very expensive military brother.

The U.S. government classifies Generation III image intensifier tubes using FOM as defined earlier in this paper. Simply multiplying tube db times lp/mm is not adequate for predicting performance in telescope systems that place tremendous demands on tube performance, due in large part to higher F numbers found in typical telescopes. In fact, too high an F number (greater than ~f20) can photon-starve the photo cathode, dramatically decreasing quantum efficiency and signal-to-noise ratio.

The bottom line for defining tube performance in telescope systems is that no single tube specification can adequately predict real-world visual performance. For example, a tube may possess a high SNR and resolution but have low PR coupled with high EBI. In such a case, a lower SNR tube with identical resolution and high PR with low EBI will visually outperform the first tube. Anyone considering ownership of a Generation III thin film intensifier should have a basic understanding of the parameters discussed in this paper.

William J. Collins
June 2004

APPENDIX 4

StellaCam II Highlights and Description*

The latest deep sky video system from Adirondack Video Astronomy is the **ASTROVID StellaCam II Astronomical CCD Video Camera.** Following in the tradition of the StellaCam Ex, the StellaCam II represents a further enhancement and advancement in Deep Sky Imaging. The images of the StellaCam II rival *cooled* conventional CCD cameras.

The StellaCam II is an excellent introduction to Deep Sky imaging for those who appreciate the versatility of a "true" video camera. The output of the camera can be displayed on large-screen televisions and monitors, including projection systems. Images can also be displayed on Laptop or desktop computers.

The on-screen (monitor) image of brighter deep sky Messier objects such as the Dumbell Nebula are equivalent to processed CCD images. The views on the monitor are spectacular…

There is a separate remote-control box for slow shutter speed (frame accumulation), gain, gamma (contrast) control, and freeze frame. The adjustment of these parameters is absolutely essential to obtain proper images.

The freeze-frame button will freeze your on-screen image on the monitor. This is useful when you have an image you want to describe or display for a large group. Once the screen image is frozen, the telescope can be slewed to the next object while a discussion is being given for the current on-screen image.

There is also PC control for remote or Internet operation of the StellaCam II.

HIGHLIGHTS

1. "True video" system for the maximum in versatility—monitor display, auto-guiding, computer image capture and image processing, etc.
2. Low noise.
3. Excellent dynamic range (allowing very faint details to be captured).

* Reprinted with kind permission of Adirondack Video Astronomy.

4. External control box for manual control of shutter speed, gain and gamma.
5. Very small size.
6. Camera runs cool, preventing, in most instances, the buildup of warm pixels.
7. Very smooth and even background with gain and gamma set at midlevel.
8. **Excellent "live" picture that is very comparable to a 10-second cooled CCD camera image.**

The StellaCam II Technology optimizes the use of the Non Ex-View SONY HAD CCD. The choice of this CCD results in an image of very low noise and very few hot pixels. The combination of the very low noise of the SONY CCD with very low noise electronics in the StellaCam II results in a marked improvement in signal-to-noise (S/N) ratio. This improved S/N ratio enables the StellaCam II to reach 18th magnitude in an eight-inch telescope. The images are very clean at midlevel gain and gamma settings. The views are so impressive compared to the visual views that it is like having a telescope 5X larger in aperture!

The 256-frame accumulation combined with internal camera electronics that are very low in noise results in an **extremely sensitive** camera. Objects like the Veil Nebula, Cocoon Nebula, and Andromeda Galaxy Dust Lanes can be easily captured.

SPECIFICATIONS:	
MODEL NUMBER	AV-STCAM-II (NTSC) AV-STCAM-II-P (PAL)
BROADCAST SYSTEM	AV-STCAM-II EIA AV-STCAM II-PCCIR
IMAGE SENSOR The camera uses the Non-Ex View version of the CCD.	AV-STCAM-I EIA SONY ICX418ALL.pdf AV-STCAM-II-P CCIR SONY ICX419ALL.pdf
CCD TOTAL PIXEL NUMBER	795 (H) × 596 (V) 811 (H) × 596(V)
SCANNING SYSTEM	525 LINES 60 FIELDS/SECOND NTSC 625 LINES 50 FIELDS/SECOND EIA
SYNC SYSTEM	INTERNAL / VD–LOCK (OPTION)
MINIMUM ILLUMINATION	**STELLACAM II 18TH MAGNITUDE DEEP SKY IMAGING WITH 8 INCH TELESCOPE** **0.00002 lux at F/1.4** Planetary – 1/60 sec ONLY (see limitations section below) Integration (Deep Sky) mode – UP to 256 Frame accumulation for a 8.5 second NTSC 10.8 second PAL exposure
RESOLUTION	600 TELEVISION LINES

GAIN CONTROL	AUTOMATIC/MANUAL GAIN 8-38dB
SIGNAL TO NOISE RATIO **MANUAL FRAME ACCUMULATION Shutter**	52 dB The OFF setting on the control box is the standard video shutter speed of 1/60 second (.017 second) for NTSC (EIA) frame and 1/50 second (.02 second) for CCIR (PAL) frame. Setting 1 is for one added frame etc....
AUTO IRIS	A.E.S./VIDEO Type Lens
CONTROL BOX–manual control of parameters	GAIN , SHUTTER SPEED GAMMA CONTROL
FREEZE FRAME	FREEZE FRAME OF IMAGE VIA FREEZE FRAME BUTTON ON CONTROL BOX.
FRAME ACCUMULATION (ON- BOARD THE CAMERA)	MANUAL CONTROL BOX – 1, 2 TO 256 FRAMES
VIDEO OUTPUT	COMPOSITE BNC
GAMMA CORRECTION	0.35, 0.45, 1.0
OPERATION TEMPERATURE	–20C TO +50C
OPERATIONAL HUMIDITY	WITHIN 85%
POWER SUPPLY	AC100V, 220V CAMERA POWER REQUIREMENTS DC10.2–13.8V 50/60Hz DC+12V Power supply weight Approx. 236g
Weight of Camera	Camera weight 150 grams

FRAME ACCUMULATION (frames added to initial frame)	NTSC	PAL
OFF	.017	.02
1	.033	.04
2	.067	.08
4	.133	.16
8	.267	.32
16	.533	.64
32	1.067	1.28
64	2.133	2.56
128	4.267	5.12
256	8.533	10.24

Index